XXXIII International Conference of Material Sciences and Its Applications (XXXIII Eg-MRS 2019)

IOP Conference Series: Materials Science and Engineering
Volume 762

Hurghada, Egypt
27 - 30 November 2019

Editors:

Arafa Hussien Aly
Numan Arshid
Amal Kasry

ISBN: 978-1-7138-2663-7
ISSN: 1757-8981

Printed from e-media with permission by:

Curran Associates, Inc.
57 Morehouse Lane
Red Hook, NY 12571

CURRAN ASSOCIATES INC. proceedings.com

Some format issues inherent in the e-media version may also appear in this print version.

Institute of Physics
Dirac House, Temple Back
Bristol BS1 6BE UK

Phone: 44 1 17 929 7481
Fax: 44 1 17 920 0979

techtracking@iop.org

Additional copies of this publication are available from:

Curran Associates, Inc.
57 Morehouse Lane
Red Hook, NY 12571 USA
Phone: 845-758-0400
Fax: 845-758-2633
Email: curran@proceedings.com
Web: www.proceedings.com

TABLE OF CONTENTS

MATERIAL SCIENCES

MATERIAL SCIENCES AND BULDING MATERIALS

IOP Conf. Series: Materials Science and Engineering **762** (2020) 011001 doi:10.1088/1757-899X/762/1/011001

Proceeding of XXXIII International Conference of Material Sciences and Its Applications (XXXIII Eg-MRS) 2019

Preface

It is with great pleasure that I introduce the proceedings of the XXXIII Eg-MRS International Conference, 27-30 November 2019, Hurghada, Egypt. This conference was dedicated to technical issue related to material sciences research and its applications.

The objectives of the conference were not only to provide opportunities to the international scientists to present state of the art research in material sciences but also to provide a forum to discuss ideas for future directions in the field and explore possibilities for collaborative research projects.

The conference program included keynote, oral, and poster presentations from scholars working in the areas of materials science and engineering from all over the world. It covered recent trends and progress made in the field of material sciences. Professors from Egypt, Germany, Malaysia, and Hungary were invited to deliver keynote lectures regarding the latest information in their respective areas of expertise.

All papers were subjected to peer-review by conference committee members and international reviewers. The papers were selected based on their quality and their relevance to the conference. The proceedings present recent advances in the fields of materials science and engineering and manufacturing as well as in other related areas including materials properties, measuring methods, analytical and numerical modeling and simulations among others. I would like to express special gratitude to members of the conference committee and organizers of the conference. I would also like to thank the reviewers for their valuable time and advice which helped in improving the quality of the papers selected for presentation at the conference and for publication in the proceedings. Finally I want to thank the authors, the members of the organizing committee, the reviewers, the chairpersons, sponsors, and all other conference participants for their support of Eg-MRS 2019.

Prof. Arafa H Aly

Beni-Suef University

Egypt

November, 2019

Conference Chairman

Prof. Dr Arafa Hussien Aly
Beni-Suef University, Egypt
Eg-MRS Chiarman

Coordinator (Eg-MRS)

Prof Dr.Hassan Talaat (Ain Shams University, Egypt)
Prof. Dr. Hassan Afify ***(National Research Center, Egypt)***
Prof. Dr.Talaat Meaz (*Tanta University, Egypt*)
Prof. Dr. Mohamed El-Zaidia (MonofiaUniversity, Egypt)
Prof Dr. Amal Kasry ***(The British University in Egypt)***

Organizing and Scientific Committee

Prof Dr. Amer El-Korashy	*(Eg-MRS)*
Prof Dr.Hesham M. Talaat	*(Eg-MRS)*
Prof Dr. Ahmed Ramadan	*(Helwan University*
Prof Dr. Sohair Negm	*(Benha University)*
Prof Dr. Hamed Hassan	*(Alexandria University)*
Prof Dr. Salah Obayya	(Zewail City)
Prof Dr. Samia A Saafan	*(Tanta University)*
Prof Dr. Gamal Turky	*(NCR Cairo)*
Prof Dr. Mohamed Shaban	(Beni-Suef Univ.)
Prof Dr. Ahmed Ghitas	(NRIAG)
Prof Dr. Soltan Eid A. Soltan	*(Helwan Uiversity)*
Dr. Afaf El-Sayed	(El Azhar University)
Dr. Mohmed Saleh	(MUST)
Ms. Aya El-Batal	(SI)

Editors and Reviewers

1- **Prof Dr.Arafa Hussien Aly, Beni-Suef Univerrsity, Egypt**
2- **Dr.Numan Arshid, State Key Laboratory os ASIC and system, Fundan University, China**
3- **Dr.Amal Kasry, The British University in Egypt**

IOP Conf. Series: Materials Science and Engineering **762** (2020) 011002 doi:10.1088/1757-899X/762/1/011002

Peer review statement

All papers published in this volume of *IOP Conference Series: Materials Science and Engineering* have been peer reviewed through processes administered by the proceedings Editors. Reviews were conducted by expert referees to the professional and scientific standards expected of a proceedings journal published by IOP Publishing.

Enhancement of Adhesion Force and Surface Conductivity of Graphene Oxide Films Using Different Solvents

F. M. El-Hossary [1], Ahmed Ghitas [2], A. M. Abd El-Rahman [3, 1], A. A. Ebnalwaled [4], Mohammed. H. Fawey [1], M. Abdelhamid Shahat [2, *]

[1] Physics Department, Faculty of Science, Sohag University, Sohag, Egypt.
[2] National Research Institute of Astronomy and Geophysics (NRIAG), Solar and Space Research Department, Helwan, Cairo, Egypt.
[3] King AbdulAziz University, Jeddah, KSA.
[4] Electronics & Nano Devices Lab, Physics Department, Faculty of Science, South Valley University, Qena, 83523 Egypt

* E-mail: m.abdelhamid999@gmail.com

Abstract. In this work, the nanotechnology procedure was utilized to improve both the adhesion force and surface properties of graphene oxide (GO) films. GO has been obtained in a powder form by oxidation purified graphite using the modified Hummer's technique. Different films of GO nanoparticles (NPs) were deposited using several types of solvents distilled water, acetone, ethanol, dimethyl formamide (DMF) or ethylene glycol. Examination of structural and optical properties of GO films were studied by XRD and UV–vis absorption spectroscopy. Moreover, electrical properties, surface roughness, contact angle, adhesion force, wetting energy and spreading coefficient were investigated. It was observed that the properties of the prepared films are influenced by the solvent type. The electrical resistivity of films is highly dependent on the solvent type which exhibited the lowest value with distilled water. Furthermore, GO film synthesized with distilled water has the best values of adhesion force and average surface roughness (Ra) 143.4 mN/m and ~ 7.83 μm, respectively. These results are mostly attributed to hydrophilic cites and GO NPs agglomeration in the surface of films and the effects of their size leading to an expansion in the surface roughness.

Keywords: Graphene Oxide, GO Thin Films, Contact Angle, Adhesion Force, Surface Roughness, Electrical Resistivity.

1. Introduction

Numerous researchers have found a great deal of enthusiasm for metal oxide semiconductors materials due to their smart benefits in engineering applications, mainly in automobiles, integrated circuits, and biological industries. [1–3]. Graphene is a significant carbon form derived from graphite and as of late, it considered as an effective semiconductor material with zero energy gap [4–6]. It is synthesized in a monolayer or a few layers of a 2D planar sheet of sp^2 hybridized carbon atoms with hexagonal structure [7,8].

Graphene and its oxide derivatives have large surface area, great electrical conductivity, dependable mechanical properties, exceptional thermal stability and good optical properties [9]. Therefore, they connected to various applications, including high-energy storage, nanomachines, electronic circuits [10], medication [11] and high-performance photovoltaics [12-14].

Graphene oxide (GO) is directly synthesized by the oxidation of purified natural graphite using a changed Hummer's technique. As reported, hydroxyl, epoxide functional groups and carboxylic acid (eCOOH) species can

be obtained during the oxidation process of graphite; all groups are responsible for the dispersion stability in polar solvents [15]. The high concentration of oxygen-enriched functional groups expand the GO interlayer spacing and in turn increase its hydrophilic nature. Therefore, GO thin films can be used in the fabrication process of polymer solar cells so as to enhance the chemical reactivity of the nanocomposite polymer layer [16-19].

As chemically known, identification of the appropriate solvent plays an important role in the synthesis process of GO films and accordingly improving their surface properties such as friction coefficient [20], optical properties [21], and thermal conductivity [22]. These surface characteristics, make it suitable material for many applications such as water desalination [23,24], gas separation [25,26], transparent conductive films [27], supercapacitors and fuel cells [28].

The mechanical and electrical characterization of industrial engineering materials can be simply explained by determining their surface wettability and surface energy [29–35]. Based on that, the wettability investigation has practical importance at the macroscopic-scale as well as at the nano-scale [36]. It leads to estimate the surface free energy, polar and dispersion energy beside the adhesion force of the film materials [37].

In this work, we concentrated on the synthesize of pure GO films with different solvent types such as distilled water, acetone, ethanol, DMF or ethylene glycol. This is to improve the electrical resistivity and surface adhesion of GO films by controlling the surface energy and attraction forces between the GO films and the glass substrate. Therefore, surface roughness, electrical conductivity, adhesion force, wetting energy and spreading coefficient were studied in order to evaluate the performance of GO films.

2. Experimental Setup

2.1. Materials

Sulfuric acid (H_2SO_4), phosphoric acid (H_3PO_4), graphite powder, potassium permanganate ($KMnO_4$, 99%) and hydrogen peroxide (H_2O_2; 30%) were used as precursors. Moreover, hydrochloric acid (HCL; 37%) and various types of a solvent such as distilled water, acetone, Ethanol, DMF and Ethylene Glycol were employed.

2.2. Synthesis of Graphene Oxide as powder form

GO in the powder form has been obtained by oxidation of the purified natural graphite using modified Hummer's technique as shown in Fig.1. In brief, three grams of graphite precursor were gradually added to 320ml of concentrated H_2SO_4 in a flask, which placed in magnetic stirrer. Then, 80ml of H_3PO_4 was added in a dropwise to the mixture under continuous stirring for 2h. After that, small portions of $KMnO_4$ (18g) were added slowly to the mixture in order to avoid fast temperature rising. The resultant combination was left for stirring for three days to allow the oxidation of graphite. The color solution changed from black to dark green. Thereafter, the H_2O_2 solution was added to terminate the oxidation process and remove the $KMnO_4$ at room temperature (RT). The resulting suspension was bright yellow in color, indicating a high level of oxidation graphite process [38, 39]. The formed graphene oxide was washed three times with (1M) of HCl aqueous solution to remove the remaining acid [40], and repeated with distilled water until a 6-7 pH was achieved. Finally, the obtained solution dried at 50°C to obtain graphene oxide powder.

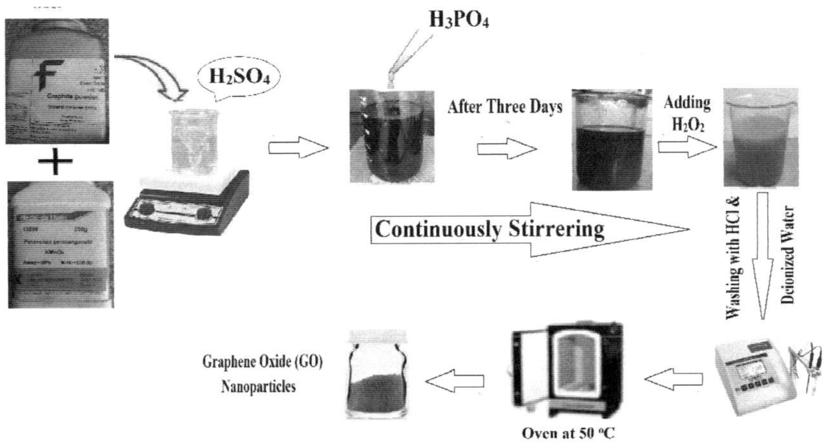

Figure 1. Preparation diagram of GO NPs

2.3. Synthesis Graphene Oxide (GO) Nanoparticles films

The clean process of the glass substrate is started using a dilute acidic solution of HCl followed by an extensive water rinse to remove the residual HCl. After that, it is sonicated in acetone from the surface of the conductive side, wet cleaned in ethanol and sonicate to remove organic impurities and finally rinse with deionized water. To deposit GO films, GO powder was added to different solutions such as distilled water, acetone, ethanol, DMF and ethylene glycol. The obtained suspensions were deposited by a spin coating on glass substrates using a 1000 rpm for 30 sec. Lastly, the obtained films were dried at 50°C.

3. Characterization

The phase purity and structure of the prepared films were characterized by X-ray diffraction (XRD) using Philips diffractometer with a Cu-target (λ=1.54056 Å) and a graphite monochromator. The optical properties of the GO films were studied using a computerized SPECORD 200 PLUS spectrophotometer with a 1 nm step, at normal incidence at RT in a wavelength range of 190-1100 nm. The contact angel analysis tool (SEO Phoenix 300) was used to measure contact angle, surface tension, wet energy, diffusion of power factor and adhesion force. Moreover, the surface roughness (Ra) was measured using the profilometer (Talysurf 50-Taylor Hopson precision). Furthermore, the four probes resistivity testing method was used to measure the resistivity by the EQ-JX2008-LD resistivity tester. All the measurements were carried out at RT (~ 26°C).

4. Results and Discussion

4.1 Structure Analysis

The X-ray diffraction pattern of a prepared GO powder reveals a strong and sharp diffraction peak (001) demonstrating a preferred orientation of GO basal planes parallel to the sample plane as shown in Fig. 2. The oxidation degree of graphite and the interlayer spacing can be acquired. The reflection peak (002) of pure graphite happened at 2θ = 25~30° with an interlayer spacing around 0.34 nm [41]. After the oxidation process (forming GO), the reflected peak is gradually weakened and finally disappeared and then appears at 2θ = 10~15° with a d-spacing value of 0.6–1.0 nm [42, 43]. This peak shift demonstrates that graphite is totally converted to GO. The main factors controlling the estimation of the d spacing value of graphite are the degree of oxidation, the removal of oxygen atoms from the graphite gallery during intercalation, and the number of water molecules involved in the interlayer spacing [42, 44].

The d-spacing of GO can be calculated according to Bragg's law:

$$n\lambda = 2d\sin\theta \qquad (1)$$

where n is the diffraction series and λ is the wavelength of the X-ray beam, d is the distance between the adjacent GO sheets or layers and θ is the diffraction angle. Herein, after oxidizing process, the native sharp peak of the as prepared GO powder occurred at $2\theta \sim 9.6^{\circ}$ and the interlayer spacing increases to about 0.92 nm. This increase in interplanar distance owing to trap oxygen functional group between the sheets in graphite structure [45]. graphite exfoliation has occurred in individual graphene oxide sheets (single or multi-layer) [46].

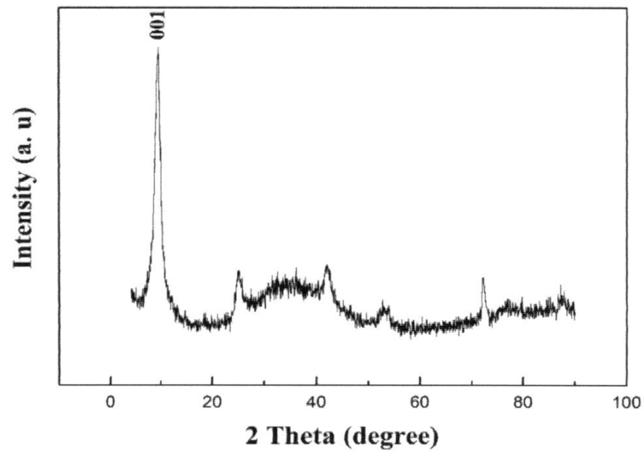

Figure 2. XRD spectra of GO film

4.2 Optical Properties

An UV-visible spectrophotometer was employed to measure the absorbance spectra of GO films, over a spectral range from 300 nm to 800 nm, which deposited using different solvents and are shown in Fig. 3. During graphite oxidation, oxygen binds to the graphene layers, which improves the layers polarity and thus improves their solubility in water [47]. These result leads to a change in the solution color from yellow to brown. It was noted that the edge of GO films uptake is about 320 nm without any change when using different solvents. On the other hand, the optical energy gap of the prepared films was determined using the universal method Tauc relationship for direct transition semiconductor materials [48].

$$\alpha h v = B\left(h v - E_g\right)^r \qquad (2)$$

Where hv is photon energy and B is a parameter which is related to the transition probability, α is the absorption coefficient, Eg is the optical band gap and r is a number that distinguishes the transition process; having a value of 1/2 for direct allowed [49]. The optical band gap values Eg were calculated and are shown in Fig. 4. Based on that, a high change is obtained for the Eg values of GO films, indicating values of 3.52, 3.56, 3.59, 3.62 and 3.67 eV for the samples deposited with several types of solvent distilled water, acetone, ethanol, DMF and ethylene glycol, respectively. Improvement of the value of the energy band gap to 3.52 eV with distilled water solvent offers high potential GO films for electronic applications.

These results may be related to the various arrangements of agglomeration epoxide functional groups and carboxylic inside GO films and the high surface roughness value in GO film which synthesized with distilled water. The change in surface roughness of the GO films affects their optical properties including transmittance, absorbance and optical constants [50, 51].

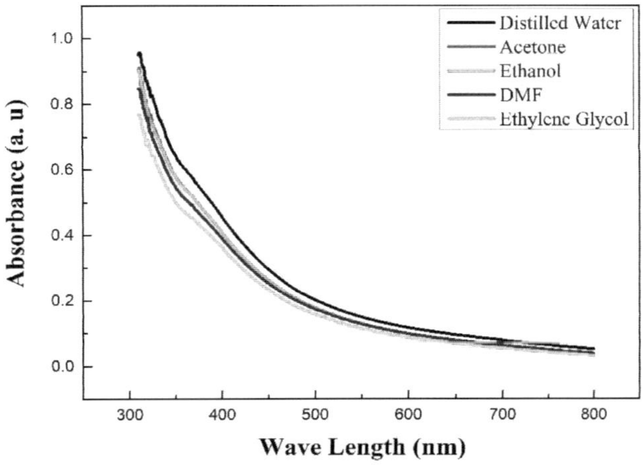

Figure 3. UV-VIS absorption spectra for GO films with different solvent types

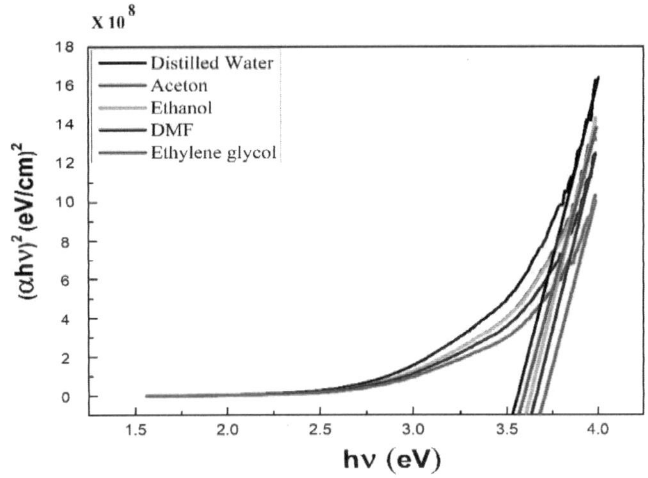

Figure 4. The relation between $(\alpha h\nu)^2$ and hν for GO films with different solvent types

4.3 Contact Angle

Contact angle (average in degree) and surface tension measurements of pure GO films, which dissolved in various solvents such as distilled water, acetone, ethanol, DMF and ethylene glycol were investigated. Contact angle value is highly depending on the utilized solvent type, which shows verities value from 14.11° to 30° for distilled water and ethylene glycol solvents as one can see from Fig. 5. additionally, the adhesion force between deposited GO films and water by calculating the work of adhesion with the equation [52]:

$$Wsl = \gamma s + \gamma l + \gamma sl \qquad (3)$$

where γs, γ*l*, and γs*l* are the solid surface free energy, liquid (water, in this study) surface free energy, and solid–liquid interfacial energy, respectively. Combining with Young-Dupré equation

$$\gamma s = \gamma sl + \gamma l \cos \theta_e \qquad (4)$$

gives the equation [53]:

$$Wsl = \gamma l\,(1 + \cos \theta_e) \qquad (5)$$

where θ_e is the equilibrium (Young's) contact angle between oxidized GO films and water, γ_L is the surface tension of water on the solid surface.

Fig .6 demonstrates that the contact angle and adhesion force of GO films are influenced by the dissolvable type. When the contact angle value was gradually increasing, the adhesion work decreased. Utilizing different solvents for synthesizing GO films leading to changes in the surface properties from hydrophobic to hydrophilic films. Using distilled water solvent for depositing GO film demonstrated to increase the hydrophilicity and has the best value of adhesion force reveals to 143.4 (mN/m).

Fig. 6 a,b represents the dependence of the spreading coefficient and wetting energy of GO films on the solvent type. The solvent type has a high effect on surface properties, leading to an improvement in the wettability of water on the GO surface. Utilizing various types of solvent for deposition GO films leads to rearrange many hydrophilic spices that in order enhance the wettability of GO films [54–57]. The arrangement of more hydrogen bonds between the used water and GO films cause an increase in the dipole/dipole interaction, which in turn improves the wettability [58, 59]. Fig. 7a demonstrates emotional improvement in the wettability that happened by utilizing distilled water as the solvent reveals to 70.6 (mN/m). In addition, the solvent type can be chosen to improve spreading coefficient GO thin films to -2.19 (mN/m) for distilled water as shown in Fig. 7b. The fundamental explanations for these outcomes are the surface roughness [60,61] and the molecular interactions [62].

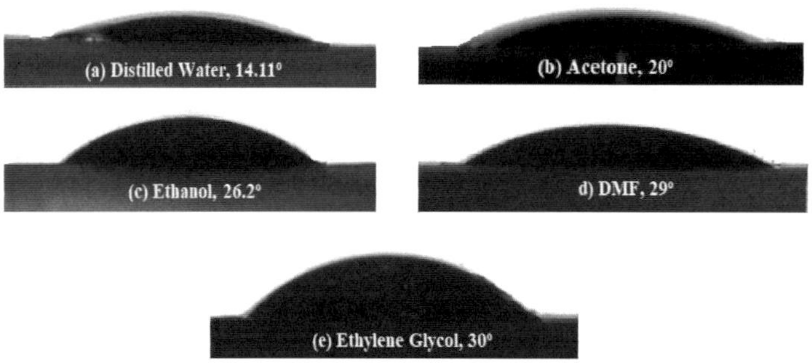

Figure 5. Contact angles and interfaces between water droplets and GO films.

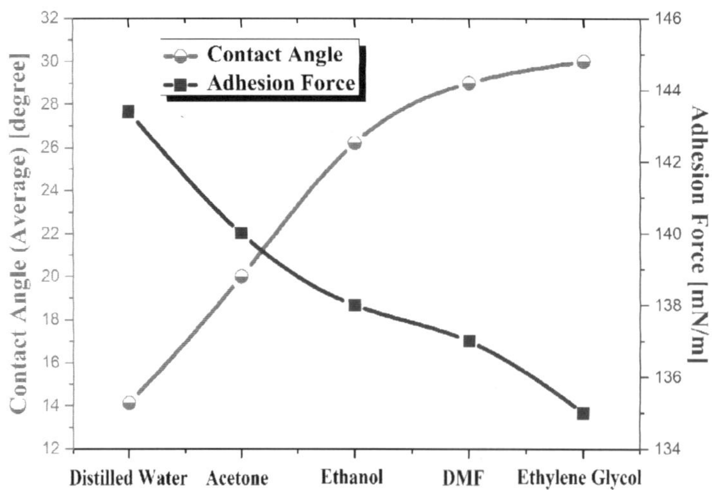

Figure 6. Effect of solvent type on the contact angle and adhesion force for GO films

IOP Conf. Series: Materials Science and Engineering **762** (2020) 012001 doi:10.1088/1757-899X/762/1/012001

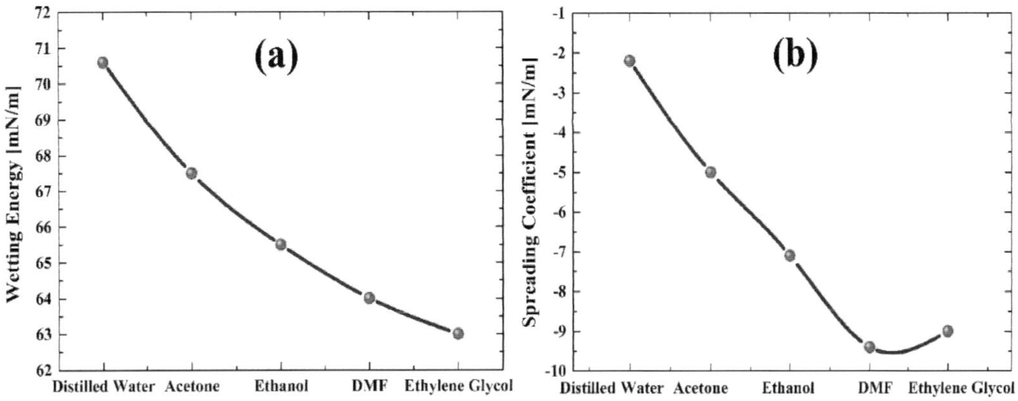

Figure 7. (a)The dependence of wetting energy, (b) spreading coefficient of GO films on solvent type

4.4 Surface Roughness Measurement

The roughness behavior is considerably different when GO films synthesized in various solvents. Fig 8. shows a relatively flat surface reduces the value of the surface roughness. The average roughness value (Ra) decreased from ~7.83 μm for distilled water solvent to ~3.75 μm for ethylene glycol. These outcomes are generally attributed to the agglomeration of GO nanoparticles in films surface and their size impacts lead to an expansion in surface roughness. However, the fibrous network structure is generally responsible for the surface smoothness. The obtained results are showing good agreement with the contact angle performances.

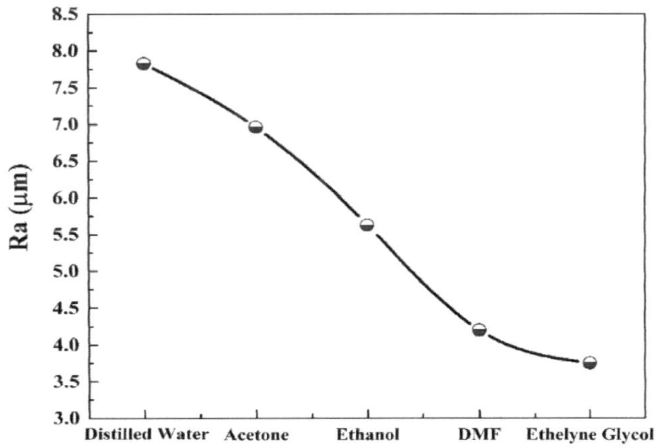

Figure 8. The average surface roughness behavior of GO films using different solvent types

4.5 Electrical Properties of GO Films

Of the generally recognized Lerf-Klinowski model for the GO structure, the basal plane of GO films contains functional groups of hydroxyl and epoxide, though the edges of GO are mainly oxidized with quinoidal and carboxylic acid species [63]. These oxygen-containing moieties give advantageous chemical handles to the expansion of other functional groups to graphene, for example alkyl gatherings, amines, and polymer chains.

Along these lines, we have been synthesized GO films using several solvent types to improve the charge carrier's mobility and reduce the electrical resistivity. The resistivity measured by 4PP method approved that; the electrical resistivity and conductivity of the obtained films are highly dependent on the types of solvents used. For the GO bulk material, the sheet resistivity determined from Van der Pauw measurements to be 1.62 MΩ/square, yielding a value of ~3.1 S/m for the bulk conductivity [64]. In the present investigation, taking into account the GO film thickness of about 100 microns, the sheet resistivity values are 0.095, 0.16, 0.1, 0.12 and 0.25 MΩ/square for the samples synthesized with solvents distilled water, acetone, ethanol, DMF and ethylene glycol, respectively as shown in Fig. 9. These obtained outcomes confirmed that distilled water solvent enhances the GO film performance and displayed an optimum value of electrical conductivity (0.11 S/m), meanwhile the existence of oxygenated groups deteriorates the electrical conduction of GO. In our case, the various electrical insulation value of synthesized GO films peppered in different solvents can be attributed mainly to the presence of a homogeneouse electrical dispersion causing various arrangements of agglomeration epoxide functional groups and carboxylic inside the GO films, and at the same time the existence of oxygenated groups [65, 66]. These result in the immobilization and obstruction of charge carriers and deteriorates the electrical conduction of GO [67,68]. Having a highly conductive GO film is synthesized in distilled water can enhance electron transfer in the framework, thus improving their electrical conductivity.

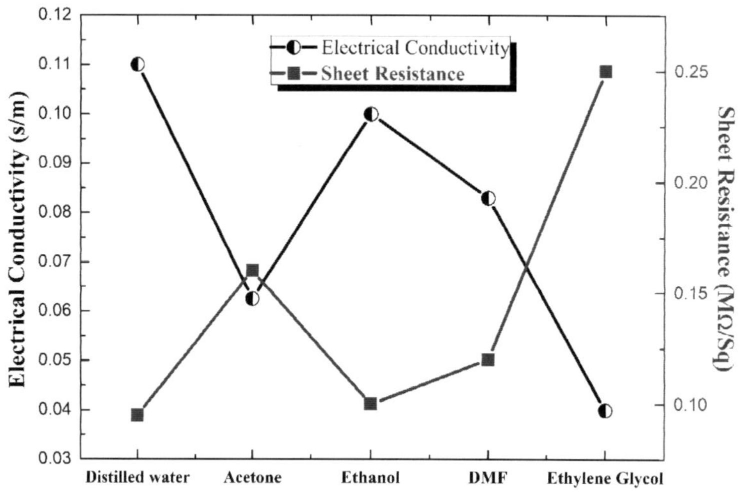

Figure 9. Electrical conductivity and sheet resistance behavior of GO films synthesized with different solvent

5. Concluding remarks

In his work, it can conclude the following points:

- Successfully synthesize metal oxide material GO NPs by oxidation purified natural graphite using modified Hummer's technique.
- GO film synthesized in distilled water solution exhibited the lowest value of electrical resistivity, and has the optimum value of adhesion force (143.4 mN/m) and average surface roughness (Ra) of ~7.83 μm.
- The obtained results are generally attributed to hydrophilic cites and the agglomeration of GO NPs in the film's surface and their size effect leads to an expansion in the surface roughness.
- The high conductivity value obtained from the GO film (0.11 S/m), which is synthesized in distilled water, can enhance the electron transfer in the framework, owing to the good arrangement of agglomeration epoxide functional groups and carboxylic inside GO film leading to improve their electrical conductivity.

References

[1] Shen, S., & Meng, Y. (2012). *Tribology Letters*, *47*(2), 273-284.
[2] Tambe, N. S., & Bhushan, B. (2004). *Nanotechnology*, *15*(11), 1561.
[3] Alfeeli, B., Cho, D., Ashraf-Khorassani, M., Taylor, L. T., & Agah, M. (2008). *Sensors and Actuators B: Chemical*, *133*(1), 24-32.
[4] J. Wei, J., Zang, Z., Zhang, Y., Wang, M., Du, J., & Tang, X. (2017). *Optics letters*, *42*(5), 911-914.
[5] Huang, H., Zhang, J., Jiang, L., & Zang, Z. (2017). *Journal of Alloys and Compounds*, *718*, 112-115.
[6] Zang, Z., Zeng, X., Wang, M., Hu, W., Liu, C., & Tang, X. (2017). *Sensors and Actuators B: Chemical*, *252*, 1179-1186.
[7] Anand, K., Singh, O., Singh, M. P., Kaur, J., & Singh, R. C. (2014). *Sensors and Actuators B: Chemical*, *195*, 409-415.
[8] Subramani, K., & Sathish, M. (2019). *Materials Letters*, *236*, 424-427.
[9] Park, J. W., Lee, C., & Jang, J. (2015). *Sensors and Actuators B: Chemical*, *208*, 532-537.
[10] Asadi, K., Timmering, E. C., Geuns, T. C., Pesquera, A., Centeno, A., Zurutuza, A., ... & de Leeuw, D. M. (2015). *ACS applied materials & interfaces*, *7*(18), 9429-9435.
[11] Shen, H., Zhang, L., Liu, M., & Zhang, Z. (2012). *Theranostics*, *2*(3), 283.
[12] Acik, M., & Darling, S. B. (2016). *Journal of Materials Chemistry A*, *4*(17), 6185-6235.
[13] Rafiee, M. A., Rafiee, J., Wang, Z., Song, H., Yu, Z. Z., & Koratkar, N. (2009). *ACS nano*, *3*(12), 3884-3890.
[14] Park, S., & Ruoff, R. S. (2009). *Nature nanotechnology*, *4*(4), 217.
[15] Sarno, M., Senatore, A., Cirillo, C., Petrone, V., & Ciambelli, P. (2014). *Journal of nanoscience and nanotechnology*, *14*(7), 4960-4968.
[16] Dreyer, D. R., Park, S., Bielawski, C. W., & Ruoff, R. S. (2010). *Chemical society reviews*, *39*(1), 228-240.
[17] Dreyer, D. R., Todd, A. D., & Bielawski, C. W. (2014). *Chemical Society Reviews*, *43*(15), 5288-5301.
[18] Tang, Q., Zhou, Z., & Chen, Z. (2013). *Nanoscale*, *5*(11), 4541-4583.
[19] Lerf, A., He, H., Forster, M., & Klinowski, J. (1998). *The Journal of Physical Chemistry B*, *102*(23), 4477-4482.
[20] Berman, D., Erdemir, A., & Sumant, A. V. (2014). *Materials Today*, *17*(1), 31-42.
[21] Kim, K. K., Reina, A., Shi, Y., Park, H., Li, L. J., Lee, Y. H., & Kong, J. (2010). *Nanotechnology*, *21*(28), 285205.
[22] Zhang, C., Hao, X. L., Wang, C. X., Wei, N., & Rabczuk, T. (2017). *Scientific reports*, *7*, 41398.
[23] Abraham, J., Vasu, K. S., Williams, C. D., Gopinadhan, K., Su, Y., Cherian, C. T., ... & Carbone, P. (2017). *Nature nanotechnology*, *12*(6), 546.
[24] Chen, L., Li, Y., Chen, L., Li, N., Dong, C., Chen, Q., ... & Zhang, L. (2018). *Chemical Engineering Journal*, *345*, 536-544.
[25] Chi, C., Wang, X., Peng, Y., Qian, Y., Hu, Z., Dong, J., & Zhao, D. (2016). *Chemistry of Materials*, *28*(9), 2921-2927.
[26] Yoo, B. M., Shin, J. E., Lee, H. D., & Park, H. B. (2017). *Current opinion in chemical engineering*, *16*, 39-47.
[27] Zheng, Q., Li, Z., Yang, J., & Kim, J. K. (2014). *Progress in Materials Science*, *64*, 200-247.
[28] Gao, W., Wu, G., Janicke, M. T., Cullen, D. A., Mukundan, R., Baldwin, J. K., ... & Dattelbaum, A. M. (2014). *Angewandte Chemie International Edition*, *53*(14), 3588-3593.
[29] Barranco, A., Borras, A., Gonzalez-Elipe, A. R., & Palmero, A. (2016). *Progress in Materials Science*, *76*, 59-153.
[30] Nieuwenhuizen, J. M., & Haanstra, H. (1966). *Philips Tech Rev*, *27*(3), 87-91.
[31] Lewis, B., & Campbell, D. S. (1967). *Journal of vacuum science and technology*, *4*(5), 209-218.
[32] Nakhodkin, N. G., & Shaldervan, A. I. (1972). *Thin Solid Films*, *10*(1), 109-122.
[33] Dirks, A. G., & Leamy, H. J. (1977). *Thin Solid Films*, *47*(3), 219-233.

[34] van Kranenburg, H., & Lodder, C. (1994). *Materials Science and Engineering: R: Reports*, *11*(7), 295-354.
[35] Kwan, J. K., & Sit, J. C. (2013). *Sensors and Actuators B: Chemical*, *181*, 715-719.
[36] Nakamura, Y., Carlson, A., Amberg, G., & Shiomi, J. (2013). *Physical Review E*, *88*(3), 033010.
[37] Torrisi, V., & Ruffino, F. (2017). *Surface and Coatings Technology*, *315*, 123-129.
[38] Kovtyukhova, N. I., Ollivier, P. J., Martin, B. R., Mallouk, T. E., Chizhik, S. A., Buzaneva, E. V., & Gorchinskiy, A. D. (1999). *Chemistry of materials*, *11*(3), 771-778.
[39] He, G., Chen, H., Zhu, J., Bei, F., Sun, X., & Wang, X. (2011). *Journal of Materials Chemistry*, *21*(38), 14631-14638.
[40] William, S., Hummers, J. R., & Offeman, R. E. (1958). *J. Am. Chem. Soc*, *80*(6), 1339-1339.
[41] Saleem, H., & Habib, A. (2016). *Journal of Alloys and Compounds*, *679*, 177-183.
[42] Ramesh, P., Bhagyalakshmi, S., & Sampath, S. (2004). *Journal of colloid and interface science*, *274*(1), 95-102.
[43] McAllister, M. J., Li, J. L., Adamson, D. H., Schniepp, H. C., Abdala, A. A., Liu, J., ... & Aksay, I. A. (2007). *Chemistry of materials*, *19*(18), 4396-4404.
[44] Van Blaaderen, A., & Kentgens, A. P. M. (1992). *Journal of Non-Crystalline Solids*, *149*(3), 161-178.
[45] Lerf, A., He, H., Forster, M., & Klinowski, J. (1998). *The Journal of Physical Chemistry B*, *102*(23), 4477-4482.
[46] Wang, H., & Hu, Y. H. (2011). *Industrial & Engineering Chemistry Research*, *50*(10), 6132-6137.
[47] Leffler, J. (2012). *Luleå University*.
[48] Pankove, J. I. (1975). *Optical processes in semiconductors*. Courier Corporation.
[49] Ahmadi, M., Ghasemi, M. R., & Rafsanjani, H. H. (2011. Journal of Materials Science and Engineering, 5, 87-93.
[50] Fang, J., Wang, J., Cao, X., Man, Y., Liu, C., Cheng, L., ... & Li, J. (2018). *Journal of Physics Communications*, *2*(1), 015009.
[51] Sreeja, S., Frobel, P. L., Mayadevi, S., Suresh, S. R., & Muneera, C. I. (2015). In *IOP Conference Series: Materials Science and Engineering* (Vol. 73, No. 1, p. 012116). IOP Publishing.
[52] Leroy, F., & Müller-Plathe, F. (2010). *The Journal of chemical physics*, *133*(4), 044110.
[53] Psarski, M., Pawlak, D., Grobelny, J., & Celichowski, G. (2019). *Applied Surface Science*, *479*, 489-498.
[54] Bronco, S., Bertoldo, M., Taburoni, E., Cepek, C., & Sancrotti, M. (2004, November). In *Macromolecular Symposia* (Vol. 218, No. 1, pp. 71-80). Weinheim: WILEY-VCH Verlag.
[55] Arpagaus, C., Rossi, A., & Von Rohr, P. R. (2005). *Applied Surface Science*, *252*(5), 1581-1595.
[56] Novák, I., & Florián, Š. (2004). *Journal of materials science*, *39*(6), 2033-2036.
[57] Novák, I., Pollak, V., & Chodak, I. (2006). *Plasma Processes and Polymers*, *3*(4-5), 355-364.
[58] Yang, L., Chen, J., Guo, Y., & Zhang, Z. (2009). *Applied Surface Science*, *255*(8), 4446-4451.
[59] Owens, D. K., & Wendt, R. C. (1969). *Journal of applied polymer science*, *13*(8), 1741-1747.
[60] Siddiqa, A. J., Chaudhury, K., & Adhikari, B. (2015). *Res. Rev. J. Med. Chem*, *1*, 43-54.
[61] Sanchis, R. M., Calvo, O., Sánchez, L., García, D., & Balart, R. (2007). *Journal of Polymer Science Part B: Polymer Physics*, *45*(17), 2390-2399.
[62] Carré, A. (2007). *Journal of Adhesion Science and Technology*, *21*(10), 961-981.
[63] Lerf, A., He, H., Forster, M., & Klinowski, J. (1998). *The Journal of Physical Chemistry B*, *102*(23), 4477-4482.
[64] Banerjee, I., Mahapatra, S. K., Pal, C., Sharma, A. K., & Ray, A. K. (2018). *Materials Research Express*, *5*(5), 056405.
[65] Liang, K., Shi, L., Zhang, J., Cheng, J., & Wang, X. (2018). *Thermochimica acta*, *664*, 1-15.
[66] Owais, M., Zhao, J., Imani, A., Wang, G., Zhang, H., & Zhang, Z. (2019). *Composites Part A: Applied Science and Manufacturing*, *117*, 11-22.
[67] Gao, Y., Gu, A., Jiao, Y., Yang, Y., Liang, G., Hu, J. T., ... & Yuan, L. (2012). *Polymers for Advanced Technologies*, *23*(5), 919-928.

[68] Huang, X., Zhi, C., Jiang, P., Golberg, D., Bando, Y., & Tanaka, T. (2013). *Advanced Functional Materials*, *23*(14), 1824-1831.

Power Saving Electrochemical Processing of Low Cost MnO_2@Porous Al Electrode for High Performance Supercapacitors Applications

M. M. Moharam, Moustafa M.S. Sanad, E.M. El-Sayed, Mai. E. Ibrahim

Chemical and Electrochemical Processing Department, Central Metallurgical Research & Development Institute, P.O. Box: 87, Helwan, Cairo, Egypt

Email: marwahaqqy@gmail.com

Abstract:

The supercapactive properties for MnO_2 coated on porous aluminum foil Al electrodes etched using various acidic baths and ethylene glycol (EG) as additive was studied. The optimum current densities for achieving high porous surface of Al foils using galvanostatic technique are 70 and 90 mA/cm^2 in case of etching from acidic solution and when ethylene glycol is added respectively. Furthermore, cyclic voltammetry, galvanostatic charge–discharge techniques and electrochemical impedance spectroscopy were performed to determine specific capacitance and cycle life of the MnO_2@ porous Al foil and MnO_2@porous Al-EG electrode using 0.5 M Na_2SO_4 as electrolyte. The cyclic voltammteric curves for the two electrodes reveal that psuedocapacitive behavior during oxidartion – reduction reaction with maximum specific capacity 221.5 F/g at scan rate 20 mV/s for MnO_2@ porous Al electrode. Moreover, the galvanostatic charge–discharge tests for MnO_2@porous Al electrode at current densities 0.5 and 1 A/g show good pseudocapacitive performance and cycle ability with nearly linear curves forming quite triangle shape. Also, The EIS curves for both electrodes elucidate that, the total impedance of MnO_2@porous Al cell is smaller than that of MnO_2@porous Al-EG cell, so, MnO_2@porous Al electrode has a high ionic conductivity between active material and ionic species. In addition to, both electrodes posse's quite similar capacitive retention about 75% after 1000 cycles and this reveals the good adhesion and less dissolution of MnO_2 film.

1. Introduction:

Recently, alternative energy resources and environmental pollution are the most important world concerns [1, 2]. In addition, the increasing demand for energy materials and improving the high-performance energy-storage devices has gained much attention [3]. Therefore, the fabrication of electrode materials has become an important research field to fabricate long cycle life, low cost, light weight and productive energy storage apparatus [3, 4]. The usage of these materials is found in various applications like solar cells [5], fuel cells [6], lithium battery [7] and supercapacitors [8] with excellent energy density and power density. Although the batteries have shared the major role in the energy storage systems for various applications ranging from wristwatches to cars and more recently electrical grid. Since, the low life cycle, low power density and high cost of energy are still the main drawbacks for batteries [8, 9]. As a result, supercapacitors are emerged as new energy storage technology in comparison to batteries due to their ability to store and release energy instantaneously, long life, lower cost and high power density. They have many applications as uses in electric vehicles, consumer electronics, portable devices and wherever a burst of energy supply is needed [2, 9, 10]. Electrochemical capacitors are commonly referred as supercapacitors, therefore, the storage mechanism of supercapacitors can be classified as electrochemical double layer capacitors (EDLCs) or non faradic supercapacitors, while the other one is faradaic pseudocapacitors or redox supercapacitors and Hybrid supercapacitors [2, 11- 15]. In hybrid SCs, both the EDLC effect and the pseudocapacitance can be simultaneously generated in a single SC. The hybrid SC one can achieve higher energy and power densities with good cycling stability by collecting Faradaic and non-Faradaic processes to store the charges.

Ruthenium oxide (RuO_2) is supposed as one of the most promising pseudocapacitors exhibiting a specific high capacitance 720 F/g [16]. However, RuO_2 is expensive and toxic which has limited its commercial viability in supercapacitors. Great efforts have been done to enhance alternative cost-effective transition metal based electroactive materials, especially, earth-abundant transition metal elements in Period 4, 5 and 6 of the periodic table as Ni, Co, Mn, Cd, Fe ,Mo, W, V and Ti, have been used to design electroactive materials because of their multivalent states [10, 17]. Obviously, some of transition metal oxides, such as Co_3O_4, NiO TiO_2, and MnO_2, also have pseudocapacitance. Manganese dioxides MnO_2 has become hotspot in recent research for its abundant resource, toxicity, wide voltage range [2, 18, 19] and good electrochemistry performance [20, 21]. Besides, the equivalent series resistance (ESR) for current collectors is considered as indicators affecting on the performances of super capacitors. The presence of the ESR in current collector has led to the reduced charge range and weakened storage capacity. Therefore, the low resistance, high conductivity, and low cost of the aluminum foils make it the acceptable choice as current collector materials, thus the etching process of the aluminum foils will greatly affect the overall performance of the supercapacitor [22]. There are three methods for etching of Aluminum foil, physical method, chemical etching, and electrochemical etching. Nowadays, the electrochemical etching is developed on the basis of chemical etching, by increasing the electric field, and fastens the process of aluminum foil etching. Therefore, the applied electric current density is proportional to the dissolution rate of aluminum foil, hence, the density of impressed current could be adjusted to control the speed of aluminum foil etching [22]. Plainly, the electro etching of Al disk was carried out using galvanostatic technique which is a simple one, easy to handle, economic and friendly environment [23, 24].

In this regard, the goal of this work is to study the anodic etching of aluminum disk using aqueous medium and ethylene glycol medium to create highly porous low cost current collector. Furthermore, the electrochemical performance of the etched Al disk as current collectors for asymmetric MnO_2 supercapacitors will be systematically envisaged. Finally, the capacitance and cycle life of the MnO_2 coating layers will be compared for the different porous Al disks anodically etched by various electrolyte systems.

2. Experimental:

2.1. Electroetching of Al disks

The highly pure Aluminum (Al disk) was etched using Galvanostatic technique to form porous Al disk. This job was done via studying various current densities ranging from (50 to 120 mA/cm^2), etching times (from 5 min to 20 min), and different electrolyte mediums. The inorganic mediums are mixture of 3 M HCl+ 1 M H_2SO_4, 3 M HCl+ 1 M H_3PO_4. Also, the organic acids as citric and acetic acid were used with the same ratio with 3 M HCl. Moreover, the effect of ethylene glycol presence as organic material with ratio 1: 4 mixture of HCl+ H_2SO_4 was studied. All of the above experiments were carried out at room temperature, while most of works related to etching of Al disk were performed at high temperatures (50-70^{0}C) [25]. All of the electrochemical experiments was performed in a standard three-electrode system using Potentiostat (Volta lab 21) at room temperature. A platinum sheet and silver/ silver chloride were served as the counter electrode and reference electrode, respectively. While, the working electrodes are Al disks. All of the reported parameters were studied to examine the optimum conditions for getting porous surface of Al disk.

2.2. Electrode fabrication

Pure MnO_2 as active material is mixed with carbon black as conductive material and binder like polyvinylidene difluoride (PVDF) in the ratio of 70:20:10 wt%. Typically, the homo-generous slurry was prepared by mixing all the components in presence of N-methyl pyrrolidone (NMP) as a solvent. Then, the aluminum disk substrates were coated using lab vacuum film coater with the prepared ink followed by drying at 70°C overnight. Then the formed electrodes (working electrodes) are ready for the electrochemical behavior measurements using battery tester.

2.3. Characterization

Cyclic voltammograms were performed with a computer controlled potentiostat (Volta-lab 21), PGP201 potentiostat, a galvanosatat 20V, and a 1A with general generator. The change in the microstructure of the formed of samples was inspected by field emission scanning electron microscopy (JEOL-JSM-5410, Japan). Moreover, lab vacuum film coating machine was used for electrode fabrication with controlled thickness around 1.55 mm. In addition to, the total porosity and pore size distribution were measured by high pressure mercury prosimeter, micrometrics 9320, USA, (30000 psi, motor driven).

2.4. Electrochemical performance measurements

The working electrodes are two types; MnO_2@porous Al disks which produced from HCl/H_2SO_4 bath and MnO_2@porous Al-EG disks which produced using HCl/H_2SO_4/EG as etching bath, while, the platinum sheet was served as counter electrode. The electrolyte solution was 0.5M Na_2SO_4 solution. Cyclic voltammetry, electrochemical impedance spectroscopy and Galvanostatic charge–discharge techniques were performed using multichannel battery tester Biologic MPG-205, France to study the electrochemical capacitive behavior and cycle life of the as-prepared electrodes.

The specific capacitance of the electrode materials can be calculated from the galvanostatic discharge curve according to the following formula [25]:

$$C = I\Delta t / m\Delta V$$

where the specific capacitance (C) applies to the single electrode (MnO_2 films), I is the discharge current. Δt and t is the discharge time of the asymmetric two electrode cell, respectively. ΔV means the voltage window of the the asymmetric device, m is the weight of the electroactive material MnO_2 on the porous Al electrode.

3. Results & discussions

3.1. Electrochemical processing of porous Al substrate

3.1.1. Effect of current density

The electrochemical etching technique includes the immersion of the Al foil in an aggressive bath to create a pitting corrosion on the surface of aluminum foil which leads to deep channels with a diameter size of few microns [26]. Figure 1 shows the potential – time transients for etching Al foil in an electrolyte solution composed of 3 M HCl:1M H_2SO_4 at various current densities starting from 50 to 120 mA/cm^2 at room temperature. It is obvious that by increasing the etching current density, the etching potential for the Al foil sharply increases during the first seconds of the anodic polarization which indicates the increasing dissolution of the metal. Then, the potential stabilizes around -0.8 to -1 V in case of 50 and 120 mA/cm^2 respectively. Chloride ion is counted one of the strongest corrosive anion, thus aluminum foil form a certain shape of corrosion hole because the halogen acid corrosion is considered to be the best medium for aluminum foil etching. The applied current density is proportional to the dissolution rate of aluminum foil. So, increasing current density from 50 to 120 mA/cm^2 fasten the process of Al foil etching [22, 26]. Figure 2 represents the FESEM images for Al foil anodized at different current densities from 50 to 120 mA/cm^2 for 15 min using 3M HCl and 1 M H_2SO_4 as electrolyte. It is found that there is no great difference in the morphology of etched Al foil through increasing current density. In addition, EDX analysis confirms that there is no formation for Al_2O_3 with raising current density from 70 mA/cm^2 to 120 mA/cm^2 during etching process and only Al metal appears at all motioned current densities as represented in Figure 3. However, the etched Al foil at 90 and 120 mA/cm^2 are so brittle and it is difficult to handle. Accordingly, the optimum current density for etching is selected at 70 mA/cm^2.

3.1.2. Effect of etching time

Figure 4 (a) reveals the FESEM image for the Al foil in absence of etching solution, while Figures 4 (b, c and d) shows the FESEM for Al foil in presence of etching solution after different etching times ranging from 5 to 15 min at 70 mA/cm ^{2} of current density. It is clear that the aluminum surface shows porous topology as expected, due to the anodic dissolution of the aluminum. Moreover, the nucleation sites are distributed homogeneously among the surface of the foil. The average roughness was found to be around 2.5–2.9 μm. Figure 5 represents the relation between weight losses of Al foil and etching time. It is found that increasing etching time enhances the dissolution of Al foil which causes thinning of Al foil. With rising etching time more than 15 min, the Al foil becomes unstable and so brittle, so it is rubbed off. For this reason, the optimum etching time is 15 min [27, 28]. In contrast to previous trials for Al foil etching using high temperature mediums [26, 27], all the above experiments are carried out at room temperature.

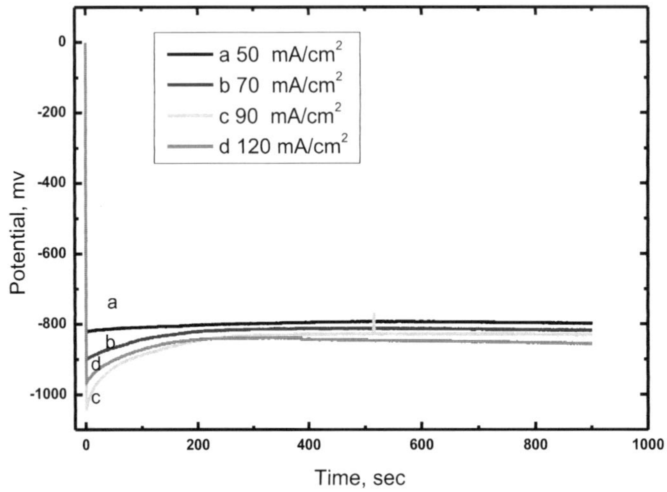

Figure 1. Galvanostatic transients for Al foil etching at various current densities for 15 min at room temperature

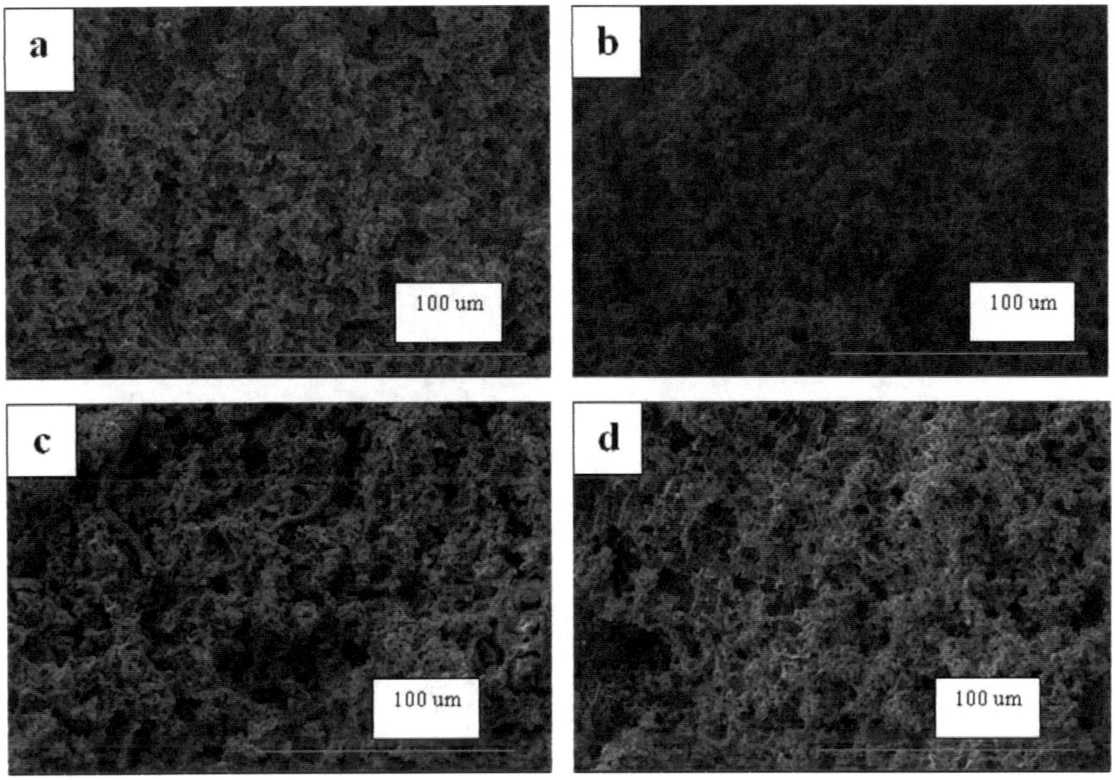

Figure 2. FESEM images of Al foil etched using H_2SO_4 in addition to HCl as electrolyte, etching time 15 min, current density, a) 50 mA/cm^2, b) 70 mA/cm^2, c) 90 mA/cm^2, d) 120 mA/cm^2

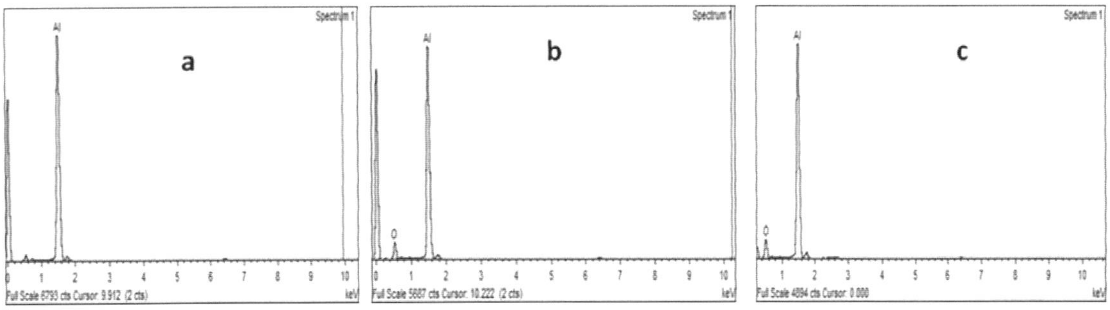

Figure 3. EDX spectrum for Al foil etched using H_2SO_4 in addition to HCl as electrolyte, current density a) 50 mA/cm^2, b) 70 mA/cm^2, c) 90 mA/cm^2 for etching time 15 min

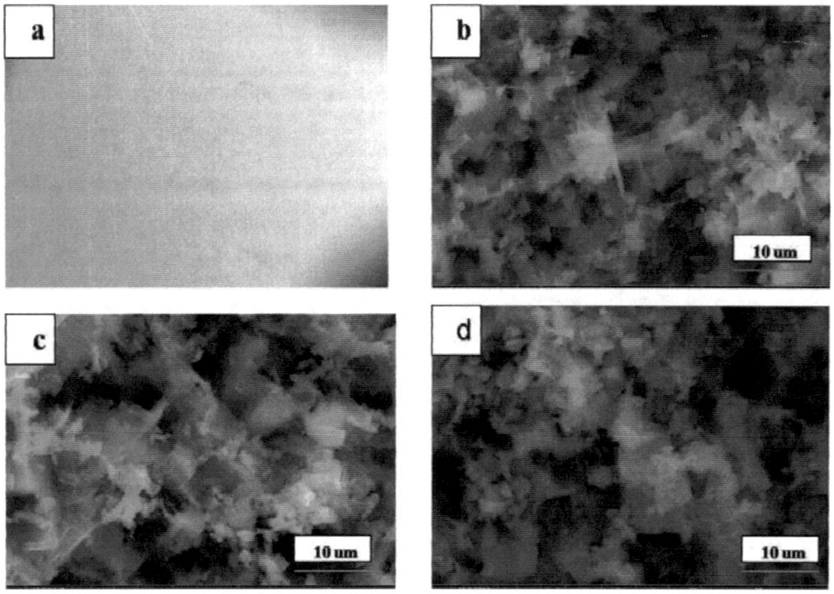

Figure 4. FESEM images of Al foil using H_2SO_4 in addition to HCl as electrolytes , current density 70 mA/cm^2, b) 5 min, c) 10 min, d) 15 min. a) blank Al foil without any treatment

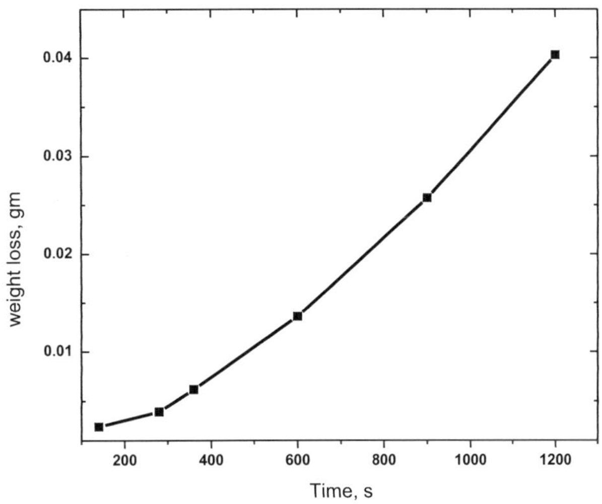

Figure 5. The relation between weight loss of Al foil and etching time at current density 70 mA/cm^2 and at room temperature.

3.1.3. Effect of anions of etching electrolyte

The FESEM images for Al foil surface using various electrolytic etching mediums such as 1 M of citric acid, 1 M oxalic acid, 1 M phosphoric acid and 1 M sulphoric acid in addition to 3 M HCl at 15 min etching time and 70 mA/cm^2 current density are appeared at Figure 6 (a, c, e, g). It is found that the nucleation sites for etched Al foils using citrate, oxalate and phosphate electrolytes are not uniformly distributed and anodic layer of Al_2O_3 is clearly observed as shown in Figure 6 (a, c & e) [29, 30]. Moreover, EDX analysis spectrum in Figs. 6 (b, d, f & h) revealed the presence of oxygen and phosphorus atoms in case of phosphoric acid. Generally, the oxide layer formation is not favorable, because it inhibits the supercapacitive property for the electroactive MnO_2 coat [29]. Furthermore, the highest weight loss for the different etching electrolytes used is significantly noticed in presence of phosphate and sulphate anions along with the HCl solution as illustrated in Figure 7. On the other hand, the surface of etched Al foil using 3 M HCl and 1 M H_2SO_4 electrolyte in Figure 6 (g) shows a uniformly distributed porous morphology and the absence of oxide layer is clearly evinced ensured via EDX analysis

IOP Conf. Series: Materials Science and Engineering **762** (2020) 012002 doi:10.1088/1757-899X/762/1/012002

as shown in Figure 6 (h). As a result, the optimum electrolyte medium for etching of Al foil in our work is a mixture of HCl and H_2SO_4 at room temperature.

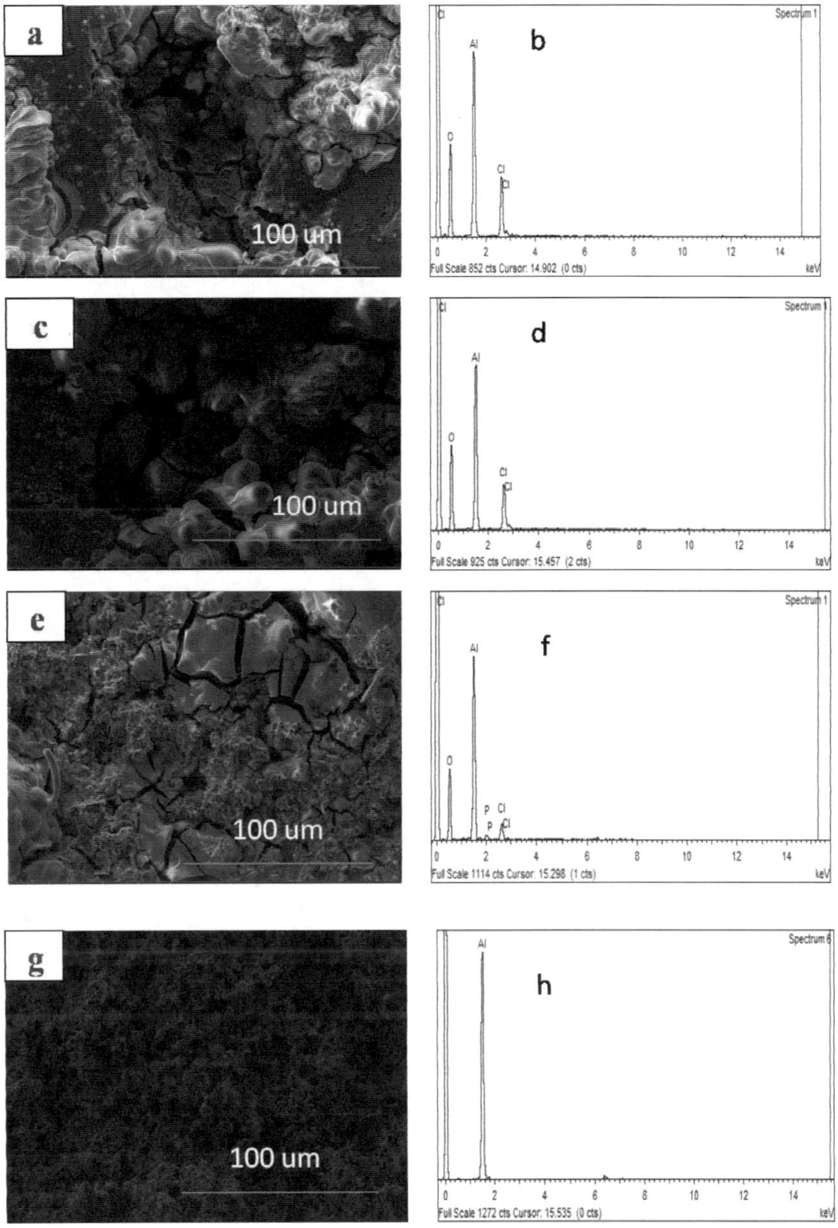

Figure 6. FESEM & EDX spectrum of Al foil using inorganic acids, a, b) citric, c, d) oxalic, e, f) phosphoric, g, h) H_2SO_4 in addition to HCl as electrolytes , current density 70 mA/cm^2, 15 min etching time.

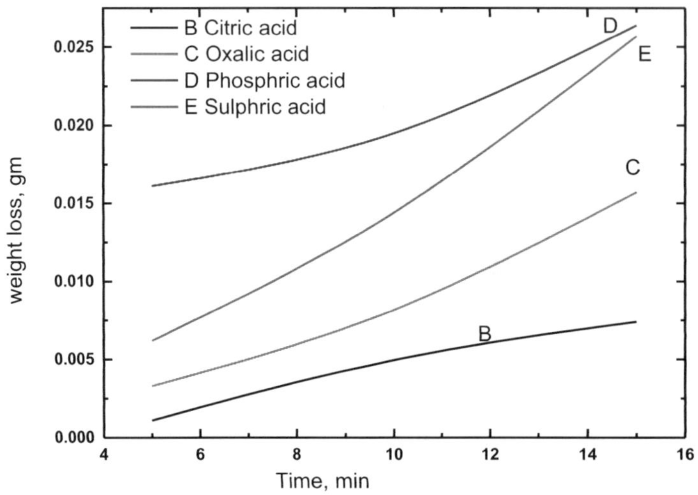

Figure 7. The relation between weight loss of Al foil and different etching electrolyte mediums used for etching process after etching time 15 min. and current density 70 mA/cm$^{2.}$

3.1.4. Effect of organic additives

Figure. 8. represents the FESEM images for the etched Al foil using ethylene glycol as organic additive in addition to the mixture of 3 M HC+ 1 M H_2SO_4 at various current densities from 50 to 90 mA/cm^2 for 15 min. as etching time. The ratio of ethylene glycol to the electrolyte mixture is 4:1 [31]. It was reported that the addition of various organic compounds such as ethylene glycol to acidic electrolytic solutions such as H_2SO_4 decreases the ability of porous anodic alumina (PAA) layer in comparison to ethanol as other organic additive [31]. In addition, the organic additives could enhance the rate of arrangement of (PAA) as a result of the gradual decrease in current density. Particularly, it was found that the addition of a miscible organic solvent with small dielectric constant e.g. ethylene glycol (dielectric constant ε= 37 at RT) [32, 33] to aqueous electrolytic solution decreases the rate constant of water which is 80.1 at room temperature [21, 31]. Figures. 8 (a-c) depict the FESEM images for the Al foil etched surfaces in presence of EG as organic additives at various current densities. It noticed

The XXXIII Eg-MRS International Conference, 27-30 November 2019, Hurghada, Egypt IOP Publishing
IOP Conf. Series: Materials Science and Engineering xxxx (2020) xxxxxxxx doi:

that more uniformity and homogeneity was achieved for the surface after addition of ethylene glycol. With increasing current density from 50 to 90 mA/cm^2, the morphology of PAA layers appeared as agglomerated layers and the average pore diameter is increased from 2.01 to 2.4 μm.

Figure 8. FESEM images of Al foil etched using ethylene glycol in addition to mixture of 3 M HCl+ 1 M H_2SO_4 as electrolyte, etching time 15 min, current density, a) 50 mA/cm^2, b) 70 mA/cm^2, 90 mA/cm^2.

3.2. Electrochemical performance of assembled electrodes

3.2.1. Specific capacitance and glavanostatic cycling

In order to evaluate the supercapacitive properties of the *MnO_2@porous Al* and *MnO_2@porous Al-EG* electrodes prepared from different etching baths, the cyclic voltammetry tests are performed within potential window 0.0 – 1.0 V at scan rates 20, 40, 60, 80 and 100 mV.s^{-1}.

Figure. 9 (a) reveals the quite rectangular-like shape for the CV curves indicating high similarity in the charge and discharge cycles of the as-prepared *MnO_2@porous Al* electrode. It is found that the maximum specific capacitance is reached about 221.5 F.g^{-1}at 20 mV.s^{-1}. In addition, the cyclic voltammetry curves of *MnO_2@porous Al-EG* electrodes prepared with addition of EG illustrate the similar psuedocapacitive behavior during the oxidation/reduction process as depicted in Figure. 9 (b). Interestingly, the estimated capacitance values are comparable with the reported value using more expensive current collector (porous Ni foam) [34-36]

Remarkably, the calculated specific capacitance values *of MnO_2@porous Al* electrodes are gradually decreased from 221.5 to 132.2 $F.g^{-1}$ with increasing scan rate from 20 $mV.s^{-1}$ to 100 $mV.s^{-1}$, respectively as shown in Figure 9 (c). Eventually, the gradual increase in the area under CV curves is a good confirmation for the direct proportionality between CV currents and scan rates of CV, which demonstrates that, an ideal capacitive behavior. Likely, the specific capacitance of MnO_2@porous Al-EG electrodes are slightly decreased from 131.5 to 45 $F.g^{-1}$ with increasing scan rate from 20 $mV.s^{-1}$ to 100 $mV.s^{-1}$, respectively as shown in Figure 9 (c). The noticeable decrease in the specific capacitance values of both type of electrodes suggests that the electrode surfaces did not have the enough exposure time to allow more cations (de-) intercalate to the bulk of the MnO_2 active material [37]. Moreover, the galvanostatic charge–discharge tests are performed on *MnO_2@porous Al electrode* in 0.5 M Na_2SO_4 at two different current density 0.5 and 1 $A.g^{-1}$ as illustrated in Figure 9 (d). The plot indicates good pseudocapacitive performance and cyclability with nearly linear curves forming quite triangle shape [37, 38].

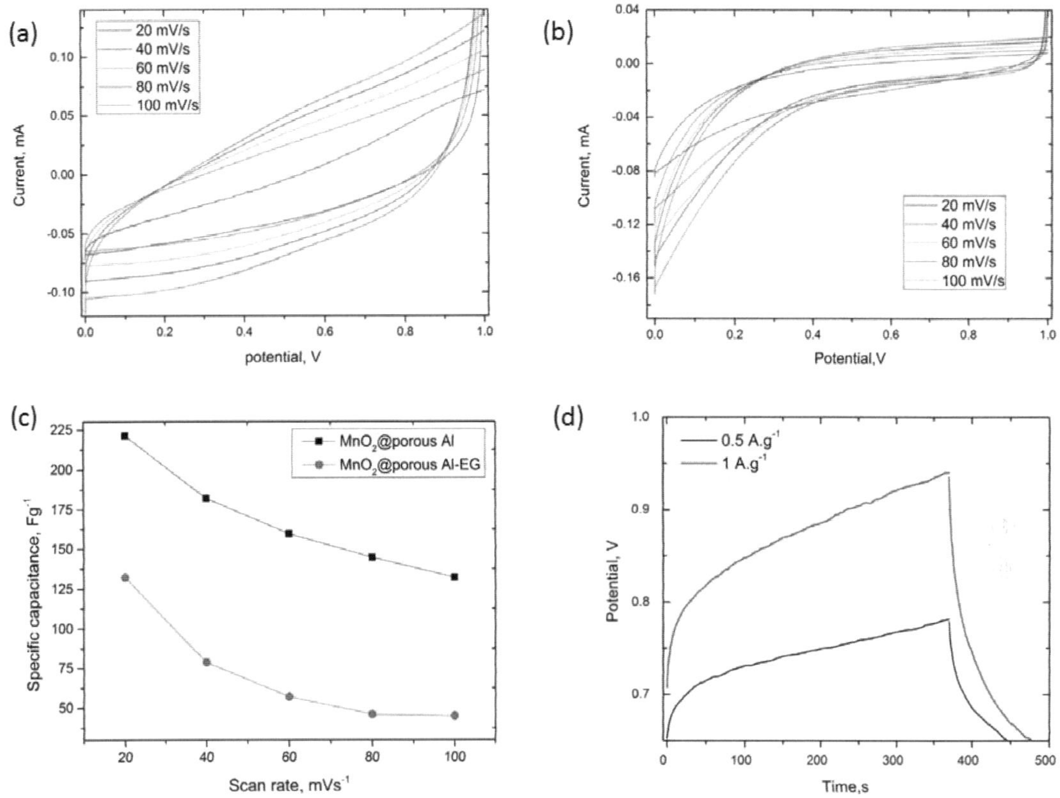

Figure 9. (a) Cyclic voltamograms of MnO_2@porous Al electrodes produced by aqueous etching with addition of ethylene glycol at different scan rates of range 20-100 $mV.s^{-1}$, (b) Cyclic voltamograms of MnO_2@porous Al-EG electrodes produced by aqueous etching with addition of ethylene glycol at different scan rates of range 20-100 $mV.s^{-1}$, (c) Variation of specific capacitance of MnO_2@porous Al electrodes with changing scan rates, (d) Galvanostatic charge-discharge curves of MnO_2@porous Al at two different current density 0.5 and 1 $A.g^{-1}$.

3.2.2. EIS studies and cycle life

Figures 10 (a and b) represent the Nyquist plots of both types of electrodes, each plot is commonly composed of at high frequency and a straight line at low frequency. This combination indicates that the electrode process was controlled by electron transport and diffusion of electronic species. Basically, the EIS spectra imply three main compartments including electrolyte resistance (R_s), charge transfer resistance (R_{ct}) and diffusion resistance of electrolytic ions (Z_w) [39, 40]. By comparing the EIS curves for both types of electrodes, it can be noticed that the total impedance of MnO_2@porous Al cell is smaller than that of MnO_2@porous Al-EG cell. Meanwhile, the slope of the straight line in the low frequency region confirms the capacitive behavior of the each electrode suggesting excellent capacitive behavior associated to inclination from 45° to 90° in the first cycle. The observed fall in inclination of the straight line after 25th and 50th cycles indicates the increase in the values of Z_w value [41].

Figure 10 (c) reveals the cycle life performance of both electrodes over 1000 cycles. It is observed that MnO_2@porous Al exhibits better cyclic performance as a result of the enhanced ionic conductivity between the active material and ionic species in the electrolyte during cycling. However, both types of electrodes possess quite similar capacitive retention about 75% after 1000 cycles which indicates good adhesion and less dissolution of MnO_2.

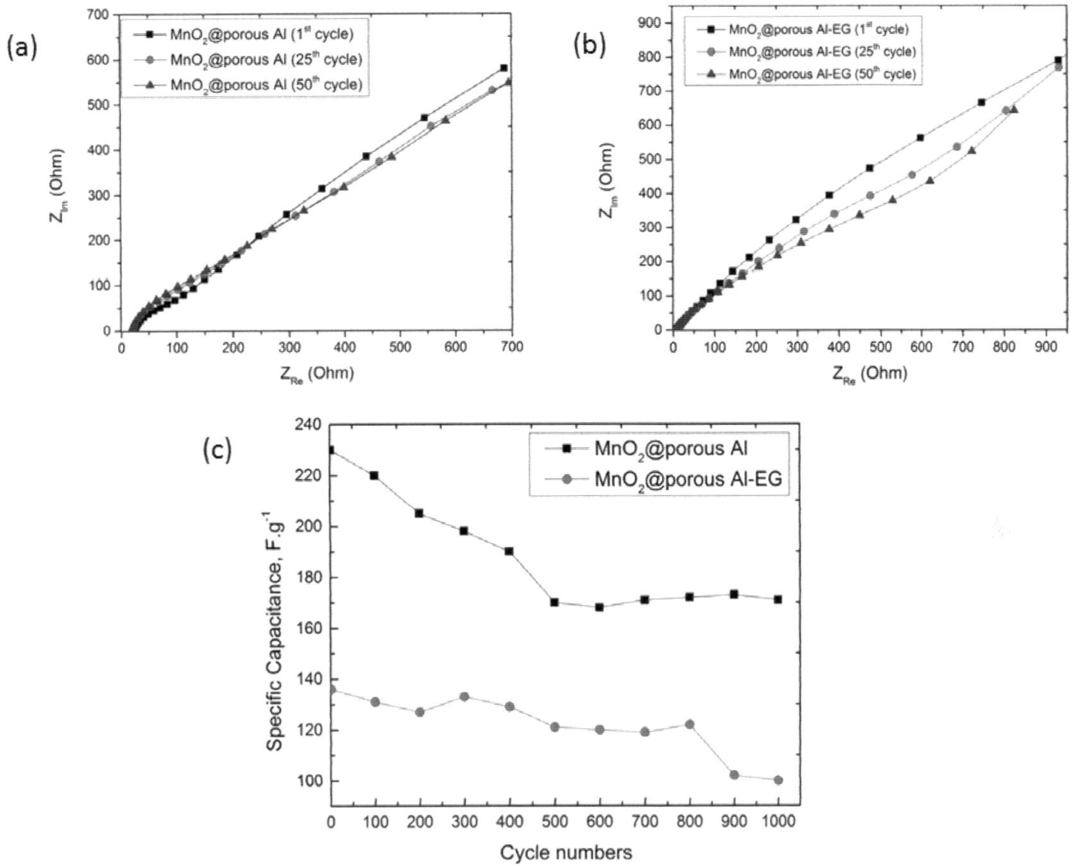

Figure 10. (a) Nyquist plots for MnO_2@porous Al cell after 1st, 25th and 50th cycles at amplitude 10 mV, (b) Nyquist plots for MnO_2@porous Al-EG cell after 1st, 25th and 50th cycles at amplitude 10 mV, (c) Cycle life performance of the supercapacitve electrodes

Conclusion:

- The supercapacitive properties for the MnO_2 @ porous Al and MnO_2 @ porous Al-Ethylene Glycol electrodes were studied.

- The etching process was carried out using various acidic mediums as citric, oxalic, phosphoric, Sulphic in addition to HCl. Also, ethylene glycol was added as organic additive.
- The mixture of 3 M HCl and 1 M H_2SO_4 is the most suitable bath for etching process, due to the absence of aluminum oxide layer and uniform distribution of porous surface as indicated by FESEM and EDX.
- Cyclic voltammetric curves for MnO_2 particles coated with screen printing technique on the porous Al foil and at porous Al-EG electrode have pseudo capacitive performance. The maximum specific capacity is recorded at 221.5 F/g at 20 mV/s for MnO2@porous Al electrode.
- The total impedance of MnO2@porous Al cell is smaller than that of MnO_2@porous Al-EG cell Using EIS data, and both electrodes have similar capacitive retention about 75% after 1000 cycles.

Acknowledgement

This work was supported and funded by Central Metallurgical Research & Development Institute (CMRDI), Cairo, Egypt.

References:

[1] Xiao . Z, Duan . T , Chen. H, Sun .K and Lu, S. *Solar Energy Materials and Solar cell* 2018 **182** 1-13.

[2] Shah. H. Ul , Wang. F , Javed M.S, Saleem . R., Nazir . M. S, Zhan. J., Khan. Z., Farooq .M. U and Ali. S, *Electrochem .Int. J.. Sci.,* 2018 **13** 6426 – 6435.

[3] Sahoo. S. , K.Naik. K , Late. D. J and Rout. C. S. , *Journal of Alloys and Compounds* 2017 **695** 154-161.

[4] Ghenaia. C and Bettayeb M, *Energy Procedia*, 2019 **159** 96-10.

[5] Shunich iFukuzumi and Joule, 2017 **1** 689-738.

[6] Zhang. H, Li. X., Liub. X.and Yan. J, *Applied Energy*, 2019 **241** 483-490.

[7] Wang J, Zhang. G, Liu. Z, Li. H, Liu. Y and Wang. Z, *Nano Energy* 2018 **44** 272–278.

[8] Zhang Q.Z, Zhang. D, Miao Z. C, Zhang X.L and Chou. S. L , *Small* 2018 **14** 1702883.
[9] Guo. Y , Bae. J, Zhao F and Yu. G, *Trends in Chemistry, doi: org/10.1016/j.trechm.* 2019 .03.005.
[10] Wang. T, Chen. H. C, Yu. F, Zhao. X.S and Wang. H, *Energy Storage Materials* 2019 **16** 545–573.
[11] Mishra R.K, , Prajapati. C.S, Shahi. R.R, Kushwaha. A.K and Sahay. P.P, *Ceramics International* 2018 **44** 5710–5718.
[12] Xu. W, Liang, Su. Y, Zhu. S, Cui. Z, Yang. X, Inoue. A, Wei. Q and Liang. C, Electrochemica acta, 2016 **211** 891-899.
[13] Dubal. D. P, Kim. J. G, Lokhande. C. D, Kim. Y, Kim. W and Holze. R, *Energy Technol.* 2014, **2**, 325 –341.
[14] Du. X, Lin. B., Li. B, Feng., Mao. S and Xu. Y. , *Adv. Funct. Mater.* 2017 **27** 1606042.
[15] Yoo. T.-H, Kim. S. M.,. Lim J. A,. Kim. J.-H,. Sang. B.-I and Song Y.-W., *J. Electrochem. Soc.* 2013 **160** ,A1772.
[16] Subramaniana. V. , Halla. Sean C. , Smithb. Patricia H. and Rambabu. B., *Solid State Ionics* 2004 **175** 511 – 515
[17] Ahmed K Yousef, Yena Kim, Piyali Bhanja, Peng Mei, Malay Pramanik, Sanad. M.M.S., MM Rashad, AY El-Sayed, Abdulmohsen Ali Alshehri, Yousef Gamaan Alghamdi, Khalid Ahmed Alzahrani, Yusuke Ide, Jianjian Lin and Yusuke Yamauchi, *RSC advances,* 2019 **9** 25240-25247.
[18] Shah. H.U, Wang. F., Javed. M.S., Shaheen. N. and Saleem. M., *Ceram. Int.*, 2017 **44** 3580.
[19] Yuan. Y, Zhan. C, He K., Chen. H, Yao. W. and Sharifi-Asl. S., *Nat. Commun.*, 2016 **7** 13374.
[20] He. W., Yang. W., Wang. C., Deng. X., Liu B., *Phys. Chem. Chem. Phys.,* 2016 **18** 15235.
[21] Juab. J., Lv Y., An. X., Liu. W., Lia. Z., Kang W.and Cheng. J., *Alloys and Compod.* 2019 **791** 1248-1256.
[22] Sujun Cheng, *CHEMICAL ENGINEERING TRANSACTIONS,* 2015 **46** 655-660
[23] Prasant Kumar Nayak and Munichandraiah . N, *Electrochemical and Solid-State Letters,* 12 2009 **6** A115-A119. ı

[24] M.Jacobl . G. , and Zhitomirsky, *Applied Surface Science*, 2008 **254** 6671-6676

[25] Shaolan Wang, Wei Li, Lipeng Xin, Ming Wu and Xiaojie Lou, *RSC Advances*, 2016 **6** 42633–42642

[26] Portet . C., Taberna. P.L., Simon. P.,and Laberty-Robert. C., *Electrochimica Acta* 2004 **49** 905–912.

[27] lei B. C., Y.dong. H, and Xin. S., *Trans. Nonferrous Met. Soc. China* 2011 **21** 133-138.

[28] Sagu J. S.., Wijayantha. K. G. U., Bohm. M., Bohm. S. and Rout. T. K., *DOI:10.1021/acsami.5b12107, ACS Appl. Mater. Interfaces* 2016 **8** 6277-6285.

[29] Ban. C.L., He. Y. d., Shao. X., Du. J., *Trans. Nonferrous Met. Soc. China* 2013 **23** 1039- 1045.

[30] Ban. C., He., wang. X. Shaom L., Mater Sci. J.: *Mater Electron* 2014 **25** 124-133.

[31] Norek. M., Stepniowski. W. J. and Siemiaszko. D., *Journal of electroanalytical chemistry*, 2016 **762** 20-28.

[32] Lu. S., Xu. D., Cui. H., Chen. Y., Jiang. L., Shen. X.and Song. Y, *Journal of The Electrochemical Society,* 2018 **165 (16)** E856-E859.

[33] Martin. J., Manzano. C.V., Caballero- calero. O., and Gonzalez. M. M. *, ACS Appl. Mater. Interfaces* 2013 **5** 72-79.

[34] Zhang. Y. J., Sun. C. and Lu P., et al., *CrystEngComm,* 2012, **14(14),** 5892–5897.

[35] Majid. S. R., *Solid State Ionics*, 2014 **262** 220–225.

[36] Kim. M., Hwang. Y. and Kim. J., J. *Power Sources*, 2013 **239** 225–233.

[37] Bhamini Bhujun, Michelle T.T. Tan, and Anandan S. Shanmugam, *Results in Physics* 2017 **7** 345–353.

[38] Xiaoyu Zhao, Yongdan Hou,b Yanfei Wang, Libin Yang, Liang Zhu, Ruge Cao and Zuoliang Sha, *RSC Adv.*, 2017 **7** 40286–40294.

[39] Ahmed Abdel-Aziz, Atef Y Shenouda, MMS Sanad, Brij Kishore, HFY Khalil, *MMB El-Sabbah, Ionics,* 2019 **25** 969-980.

[40] Sanad. M.M.S., Rashad. M.M., and Shenouda., Atef Y. , *Int. J. Electrochem. Sci,* 2016 **11** 4337-4351.

[41] Moustafa M.S. Sanad, *Materials Research Express*, 2019 **6** 105508.

IOP Conf. Series: Materials Science and Engineering **762** (2020) 012003 doi:10.1088/1757-899X/762/1/012003

Optimization of Spin Coated TiO_2 Layer for Hole-Free Perovskite Solar Cell

Hassan Elshimy*, Tamer Abdallah and Ali Abou Shama

Physics Department, Faculty of Science, Ain Shams University, Cairo, Egypt
**Email:* hmh_elshimy@hotmail.com

Abstract. The optimization of the TiO_2 layer in free-hole perovskite solar cells is critical, as it strongly affects the device performance. In this study, simple FTO/TiO_2/ $CH_3NH_3PbI_3$/C solar cells were fabricated, via two-step dipping method, with different thickness of TiO_2 layer. The structure and microstructure of the TiO_2 layers were investigated by X-Ray diffraction (XRD), Raman spectroscopy, and UV spectroscopy (UV-vis). The effect of different TiO_2 thickness on the structure and photovoltaic performance of the fabricated devices were investigated using XRD and IV measurement, respectively. The photovoltaic parameters, including short-circuit-current (J_{sc}), open-circuit-voltage (V_{oc}), Fill factor (F.F) and power conversion efficiency (PCE), were found to be strongly dependent on the thickness of TiO_2 layer. The investigation showed that, efficient device can be fabricated by optimizing the TiO_2 thickness, employing a thickness of 108 nm as an optimum thickness in this study.

Keywords: Perovskite, $CH_3NH_3PbI_3$, TiO_2, Solar cell.

1. Introduction

Recently, the organic–inorganic perovskite $CH_3NH_3PbI_3$ has attracted new and promising class of materials for photovoltaic applications due to their large absorption coefficient, high carrier mobility, longer free charge carrier lifetimes and broad range of light absorption [1–11]. The advantages of low cost, facile fabrication and high-power conversion efficiency (PCE) make perovskite solar cells (PSCs) competitive with conventional silicon based solar cells [12,13]. Typically, PSCs are composed of five parts: conductive glass (FTO or ITO), electron transport layer (ETL), a perovskite absorber layer, hole transport layer (HTL) and counter electrode (Au, Ag or C).

The electron transport layer (ETL) is indispensable and plays an important role in the performance of PSCs. It is responsible for extracting the photogenerated electron from the perovskite layer and transporting these electrons to the external circuit. In addition, prevent the holes generated in the perovskite from reaching the conductive glass to avoid short circuit [14–16]. Therefore, high transparency, good charge collection capability, suitable bandgap and appropriate thickness of ETL contribute significantly to the reduction of carrier recombination and hence the performance of the PSCs.

Several organic and inorganic materials have been used as electron transport layers, such as ZnO, SnO_2, $CsCO_3$, PCBM and TiO_2 [17–22]. Among these materials, TiO_2 is the most widely used as an ETL because of its environmentally friendly nature, wide band gap, high electron mobility and good stability [23–26]. The compact TiO_2 layer can be produced either by spin coating, spray pyrolysis, atomic layer deposition (ALD), electrochemical deposition or thermal oxidation [8,27–33]. Among them, spin coating is considered as the simplest and more cost-effective method for depositing TiO_2 layers, where a TiO_2 precursor solution is spread onto a FTO substrate and spun to form a thin transparent film. In particular, the crystallographic form, surface modification, temperature treatment and the film thickness of the TiO_2 layer have a great influence on the crystallinity and thickness of the perovskite layer and hence the performance of the PSC [16,34–36].

Herein, it is interesting to investigate the effect of TiO_2 thicknesses (using spin coating method) on subsequently deposited perovskite film and the photovoltaic performance of the PSCs. In the aim of simplifying the cell structure besides reducing the manufacturing cost, hole- free perovskite solar cells were fabricated using a commercial carbon paste, as a counter electrode, instead of Au and Ag.

2. Materials and methods

2.1. TiO_2 samples preparation

Initially, the fluorine-doped tin oxide (FTO) glass substrates were etched with HCL and Zinc powder. The substrates were ultrasonically cleaned for 10 minutes in each of: acetone, isopropanol and de-ionized water, to remove any traces of impurities, then air dried.

To prepare the TiO_2 precursor solution, a solution consisting of 175 µL of titanium (IV) isopropoxide in 1.25 mL of anhydrous isopropanol was added to a solution of 17.5 mL of HCl in 1.25 mL of anhydrous isopropanol [37].

The spin-coating of the TiO_2 solution was carried out in air at 3500 rpm for 30 s and consequently dried at 240 °C for 1 min. The thickness of the TiO_2 layer was tuned by adjusting the number of the above spinning- drying cycle. Subsequently, the resulting films were annealed in air at 450 °C for 45 min. The films with 1, 3, 6, 9, and 12 spin-coated layers were denoted 1X, 3X, 6X, 9X and 12X, respectively.

2.2. Device fabrication

The two-step solution deposition, as a facile and low-cost method, was used to fabricate the perovskite cells [8]. PbI_2 solution (1 M) was prepared by dissolving 462 mg PbI_2 in 1 ml N,N-dimethylformamide under stirring at 70 °C. The different TiO_2 films were infiltrated with PbI_2 by spin-coating a PbI_2 solution at 6000 rpm for 30 s. The obtained films were then annealed at 70 °C for 30 min. After cooling to room temperature, the films were dipped in a solution of CH_3NH_3I in isopropanol (10 mg ml^{-1}) for 20 s and rinsed with isopropanol. Finally, once the perovskite film cooled down to room temperature, the commercial carbon paste was deposited on the films by doctor–blading technique and dried at 100 °C for 30 min.

2.3. Characterization

X-ray diffraction patterns of the samples were collected on a Philips diffractometer (X'pert MPD). The diffracted intensities were collected using Cu K_α radiation of a copper sealed tube charged with 40 kV voltage and 30 mA. To correct instrumental broadening, LaB6 standard was used. The crystalline structure of the TiO_2 film was studied using Raman spectroscopy by using HORIBA JOBIN YVON Ramanor HG2S spectrometer, equipped with a micro-Raman attachment and Ar+ laser excitation (514.5 nm).

The optical properties of TiO_2 films were studied using ultraviolet-visible spectroscopy (UV–vis). The diffuse reflectance spectra of the TiO_2 thin films, were obtained using a Shimadzu UV-2700 UV–vis spectrophotometer in the wavelength range from 300 nm to 800 nm.

The current density-voltage (J-V) characteristics of PSC devices were measured in air with a digital source meter (Keithley 2400) under one-sun illumination (AM 1.5 G, 100 mW cm^2 at 25 °C) from a solar simulator (Sol3A Class AAA). The cross-section morphology of perovskite cell was investigated by using a scanning electron microscope (SEM, Quanta FEG 250) at 20 kV.

3. Results and Discussion

XRD pattern of the Glass/TiO_2 thin film is shown in Figure 1. The phase identification ratifies the formation of a Tetragonal anatase TiO_2, (ICDD: 01-089-4921). The inset shows the main peak, (101), for the TiO_2 films of different layers. As shown, the peak intensity is enhanced as the number of layers increased; which is an indicator for increasing the film thickness. This is due to the fact that the diffracted intensity is highly affected by the film thickness. Moreover, increasing the number of layers lead to a slight reduction in the integral breadth which gave rise to an increase in the crystallite size. The average crystallite sizes of TiO_2 particles for the different films were estimated using Scherrer formula, given by

$$D = \frac{K\lambda}{\beta \cos\theta},$$

where D is the crystallite size, *k* is a constant, λ is the x-ray wavelength (0.1540 nm), β is the integral breadth of the diffraction peak. For more realistic calculation, the broadening of the glass substrate was deconvoluted. The calculations estimated an average crystallite size of 8, 11, 13, 14 and 17 nm±2 for the film coated by one layer (1X), three layers (3X), six layers (6X), nine layers (9X) and twelve layers (12X), respectively.

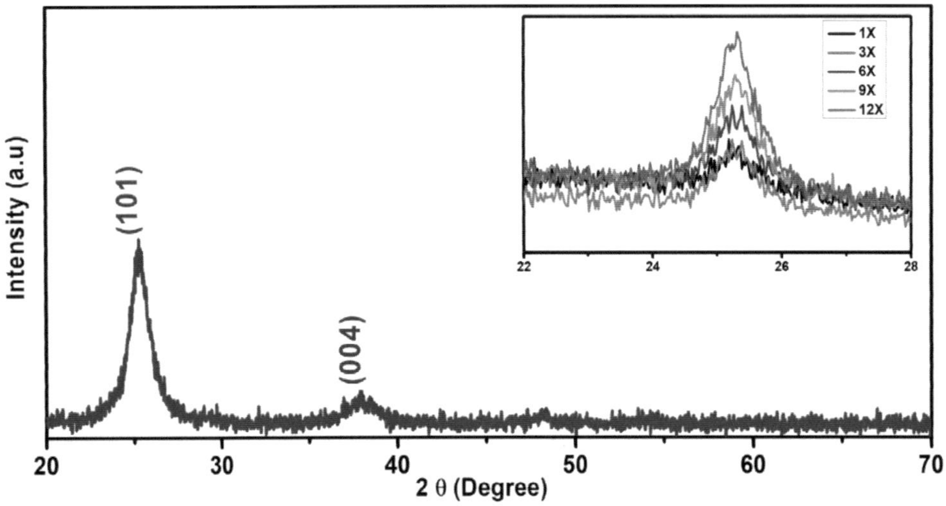

Figure 1. The XRD of TiO_2 film. Inset shows the XRD of (101) plane for different TiO_2 films (The number before "X" in the legend indicates the number of TiO_2 layers spin coated on the substrate)

In order to confirm the anatase phase for the TiO_2 film, Raman spectroscopy was performed and a Raman bands at 398, 517, 640 and 798 cm^{-1} were obtained, Figure 2. The obtained bands are a characteristic bands for the anatase phase. The bands at 640 cm^{-1} is assigned to the E_g mode and the band at 398 cm^{-1} to the B_{1g} mode. The band at 517 cm^{-1} is a doublet of the A_{1g} and B_{1g} modes. The weak band at 798 cm^{-1} is the first overtone of the B_{1g}

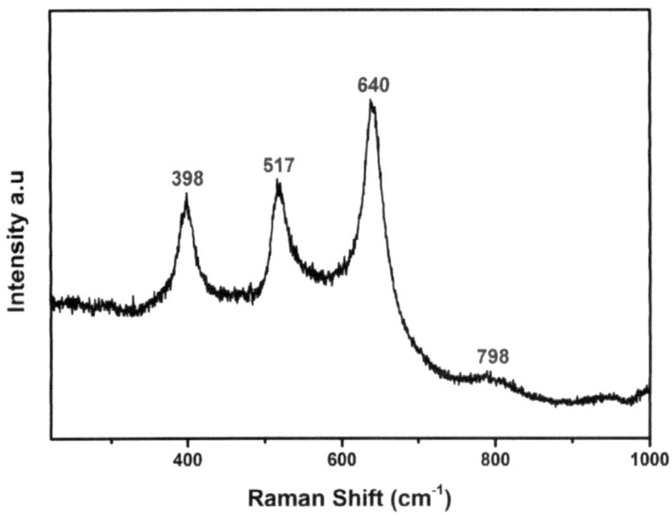

Figure 2. Raman Spectrum of TiO_2 film. (One-layer TiO_2 film,1X)

mode.

Figure 3 shows the variation of the optical absorbance versus wavelength for the TiO_2 thin film samples, in the wavelength range between 200 nm and 800 nm. The graphs show that the absorbance decreased sharply in the 300–350 nm wavelength range and then become relatively constant. From the plot, it is obvious that the absorbance of the films goes on increasing with increment the number of spin coating cycle owing to increased film thickness. Furthermore, the absorption edge shifts towards longer wavelengths indicating a decrease in the band gap as a result of increasing the crystallite size.

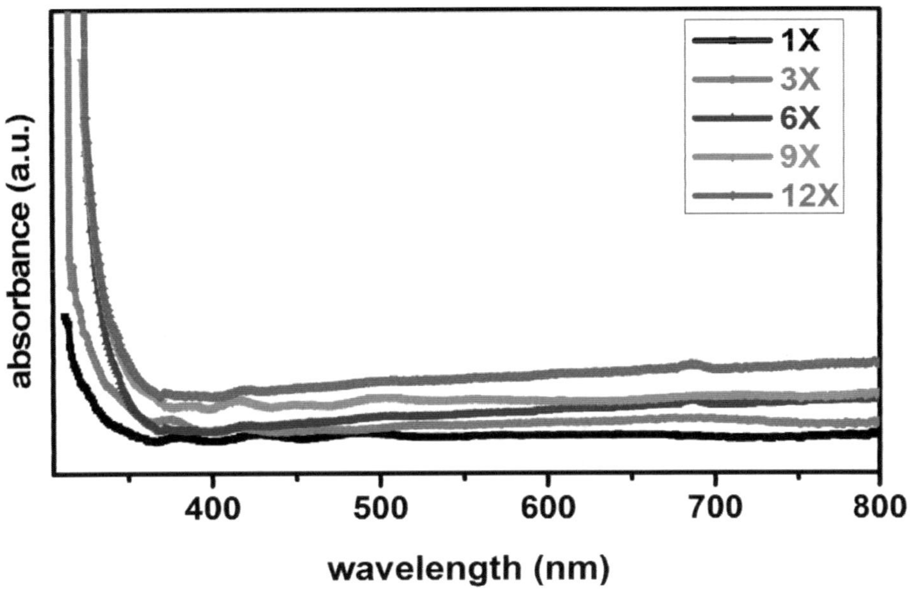

Figure 3. The optical absorbance for TiO_2 thin film samples.

The most common method to calculate the optical band gap for the thin films is fitting the data to Tauc's relation of the form [38],

$$\alpha h\nu = \beta\left(h\nu - E_g\right)^n$$

where, $h\nu$ is the incident photon energy, E_g is the optical band gap, β is a constant that depends on the transition probability and α is the absorption coefficient which is computed from the absorbance(A) using the equation , $\alpha(\lambda) = 2.303\ A(\lambda)/t$ (t is the film thickness). The exponent n takes different values depending on the mechanism transitions, which is 1/2 for direct and 2 for indirect allowed transitions.

Absorption spectrum fitting (ASF) method is another powerful method used to determine the optical band gap [39,40]. This method depends on the variation of the optical absorbance versus wavelength of the material only without the need of the film thickness. In this procedure, Tauc's relation can expressed as a function of the wavelength (λ):

$$\frac{A(\lambda)}{\lambda} = C\left(\frac{1}{\lambda} - \frac{1}{\lambda_g}\right)^n \quad \text{Where } C = \beta\left(h\,c\right)^{n-1}\left(\frac{t}{2.303}\right)$$

Here, h, c and λ_g are Planck's constant, the velocity of light and the wavelength corresponding to the optical bandgap, respectively. This equation could be used to estimate the optical band gap without the need of the film thickness. The value of the band gap in eV can be calculated from the parameter λ_g using

$$E_g^{ASF} = 1240/\lambda_g$$

The value of λ_g can be determined by extrapolating the linear part of $\left(\frac{A}{\lambda}\right)^{1/n}$ versus $\left(\frac{1}{\lambda}\right)$ curve at $\left(\frac{A}{\lambda}\right)^{1/n} = 0$.

Figure 4 shows the ASF plot for the films coated once, 1X, and 3X, 6X, 9X and 12x. As a benefit of ASF method, the slope of the straight line for the linear region is a good indicator for the change of the film thickness. It is observed that, as the number of coating cycle increases, the slope increases, indicating an increase in the film thickness, Figure 4 (f).

Using the fitting parameters, the band gap energies for the film coated by one layer(1X), three layers (3X), six layers(6X), nine layers(9X) and twelve layers(12X) are 3.28, 3.26, 3.25, 3.25 and 3.25 eV, respectively. It is obvious that, with the increase in the film thickness, band gap slightly decreases as a consequence of increasing the crystallite size.

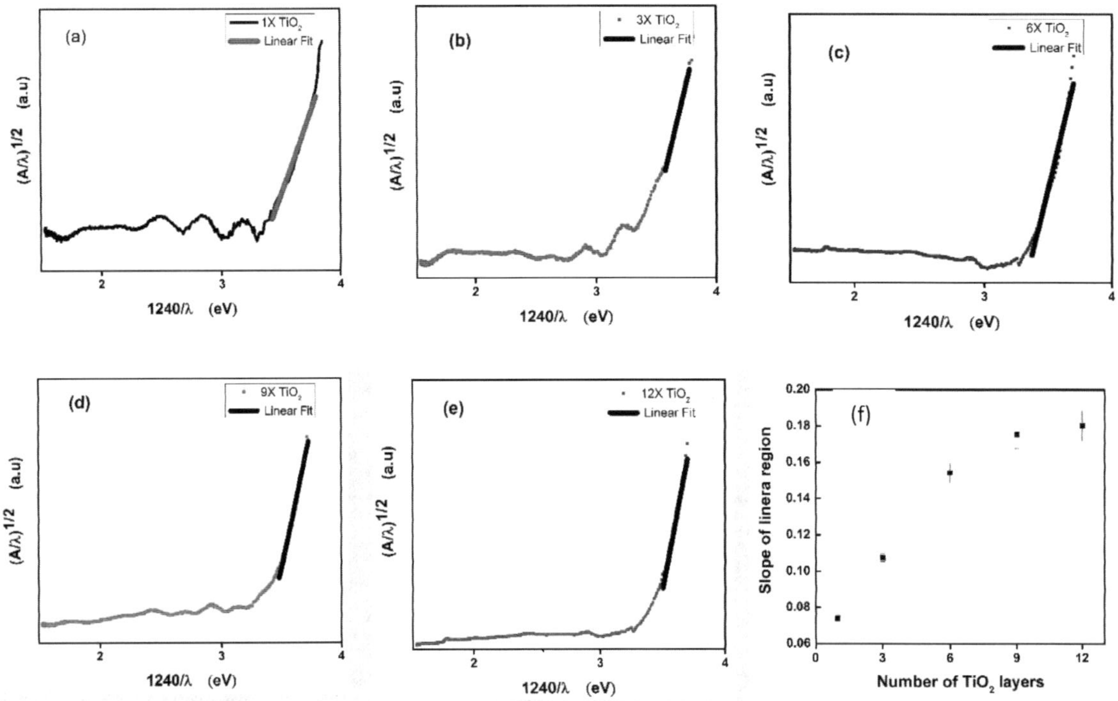

Figure 4. ASF plot for the TiO_2 film spin coated by a) one layer(1X) , b) three layers (3X) , c) six layers(6X) , d) nine layers(9X) and e) twelve layers(12X) . f) Slope as a function of number of layers

The XRD patterns for $CH_3NH_3PbI_3$ based on one (1X), three (3X), six (6X), nine (9X) and twelve (12X) TiO_2 layers are displayed in Figure 5. The $CH_3NH_3PbI_3$ films gives diffraction peaks at 14.14, 20.05, 23.53, 24.53, 28.50, 31.95, 40.54 and 43.11°, which are assigned to the (110), (020), (121), (022), (220), (130), (224) and (134) planes of the tetragonal (I 4 c m) perovskite structure with a = 8.85 Å and c = 12.63 Å. In addition, an additional diffraction peaks at 26.48, 33.74 and 37.80° which are assigned to the (110), (101) and (200) planes of the tetragonal SnO_2. Besides a peak at 12.66° which is belong to the (001) plane of the Hexagonal PbI_2.

Figure 6 shows the dependence of the diffraction intensities of both (110) plane of $CH_3NH_3PbI_3$ and (001) plane of PbI_2 on the TiO_2 thickness. It is obvious that, by increasing the TiO_2 thickness, the diffraction intensity of both (110) plane of $CH_3NH_3PbI_3$ and (001) plane of PbI_2 increases. The increase in the $CH_3NH_3PbI_3$ diffraction intensities might attributed to the thicker TiO_2 which give rise to more PbI_2 absorbed and therefore thicker $CH_3NH_3PbI_3$ film formed. Meanwhile, the ratio of PbI_2 to $CH_3NH_3PbI_3$ intensities increases, which is indication for increasing the amount of un-reacted PbI_2.

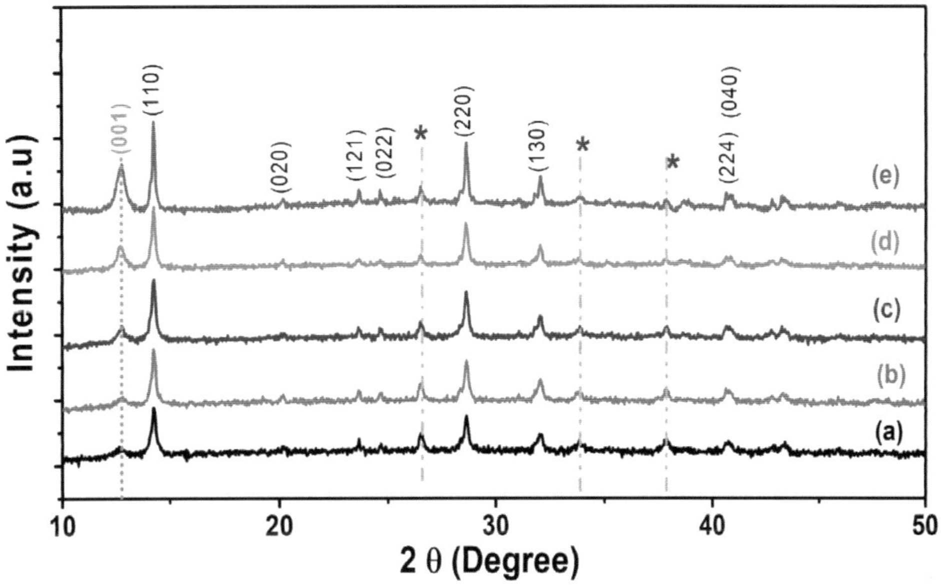

Figure 5. The XRD patterns for $CH_3NH_3PbI_3$ based on different TiO2 layers. (a) one layer (1X), (b) three

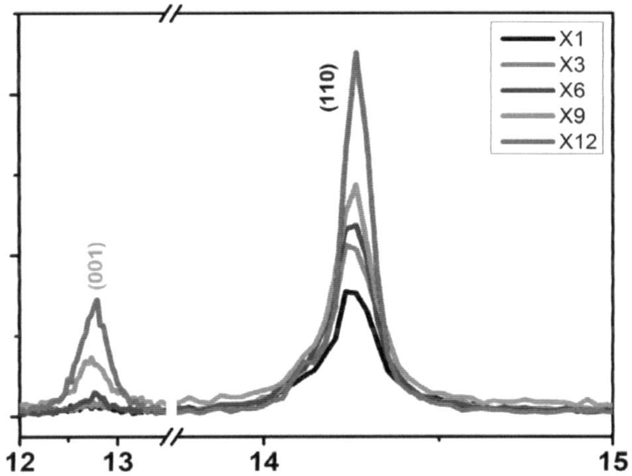

Figure 6. Diffraction peaks for (001) of PbI_2 and (110) of $CH_3NH_3PbI_3$ for different TiO_2 based films

The typical J-V curve of the best-performing devices based on one(1X), three(3X), six(6X), nine(9X) and twelve(12X) TiO_2 layers are shown in Figure 7. In order to ensure the reliability of our results, for each TiO_2 thickness, we fabricated 3 devices each containing 3 cells and tested under the same conditions. The corresponding average photovoltaic parameters, including open-circuit potential (V_{oc}), short-current density (J_{sc}), Fill factor (FF) and PCEs, are shown in Figure 8

The average short-circuits current for PSCs with one (1X), three (3X), six (6X), nine (9X) and twelve (12X) TiO_2 layers are 4.785 mA/cm^2, 6.465 mA/cm^2, 7.2 mA/cm^2, 5.7 mA/cm^2 and 5 mA/cm^2, respectively. While the open- circuit voltage are 0.64 V, 0.8 V, 0.82 V, 0.75 V and 0.64 V, respectively.

We found that the J_{sc} increased with increased the number of TiO_2 layers until 6 layers but decreased for devices with nine and twelve TiO_2 layers. For the devices with one, three and six TiO_2 layer, the increase in the J_{sc} might attributed to the increase of the surface roughness which give rise to enhance the interfacial area and therefore the rate of electron transfer. Moreover, increasing number of TiO_2 layers may result in absorbing more PbI_2 and therefore a relative thicker $CH_3NH_3PbI_3$ is formed, as demonstrated by XRD.

Upon increasing the number of TiO_2 layers, nine and twelve layers, the J_{sc} decreased, we speculate that might be due to the effect of charge accumulation in the TiO_2/ $CH_3NH_3PbI_3$ interface that obstructs the interfacial electron transfer on the surface of the TiO_2 layer, as result of the un-reacted PbI_2. In addition, the conductivity of TiO_2 might be a problem for a thicker film.

The trend of the average open circuit voltage (V_{oc}) and the average fill factor (FF) is similar to that of the average short circuit current (J_{sc}). The drop in the open circuit voltage can be explained based on the classic theory model [41,42], the V_{oc} is related to the ETL thickness by

$$V_{oc} = -A \ln d + B$$

Where d is approximately the average thickness of the (ETL) and the terms A and B are depending on the absorption Coefficient, thickness, charge generation efficiency, thermal activation energy, recombination constant of the perovskite film, etc. Assuming A and B are constant, which don't vary with the TiO_2 thickness, V_{oc} is inversely proportional to the TiO_2 thickness. Furthermore, the illuminated light would be scattered strongly by the thick film, resulting in a great loss of light energy when passing through the film, leading to decrease the average photogenerated carrier and hence the V_{oc} [42].

Figure 8(b) show that, the devices with six TiO_2 layers exhibit the best PCE among the five devices, 2.55%, with V_{oc} (0.82 V), J_{sc} (7.2 mA/cm^2) and FF (43.2%). The best performance is attributed to good surface roughness, thicker $CH_3NH_3PbI_3$ layer and low recombination rate.

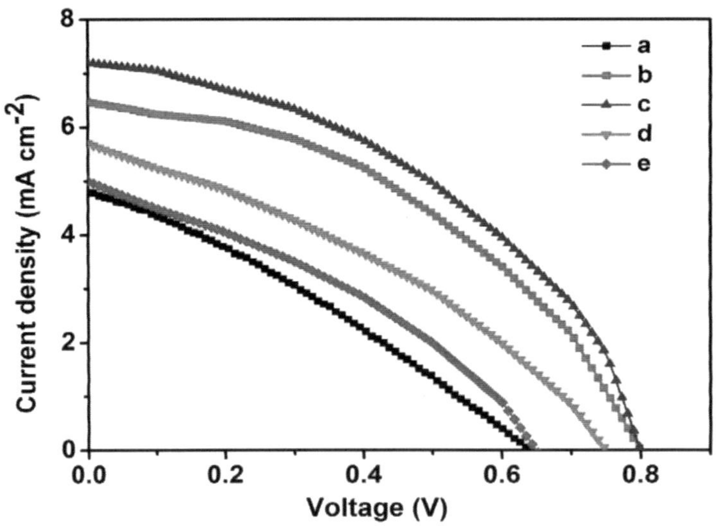

Figure 7. J-V curves of the PSCs based on different TiO2 thicknesses: (a) one layer, (b) three layers, (c) six layers, (d) nine layers and (e) twelve layers.

IOP Conf. Series: Materials Science and Engineering **762** (2020) 012003 doi:10.1088/1757-899X/762/1/012003

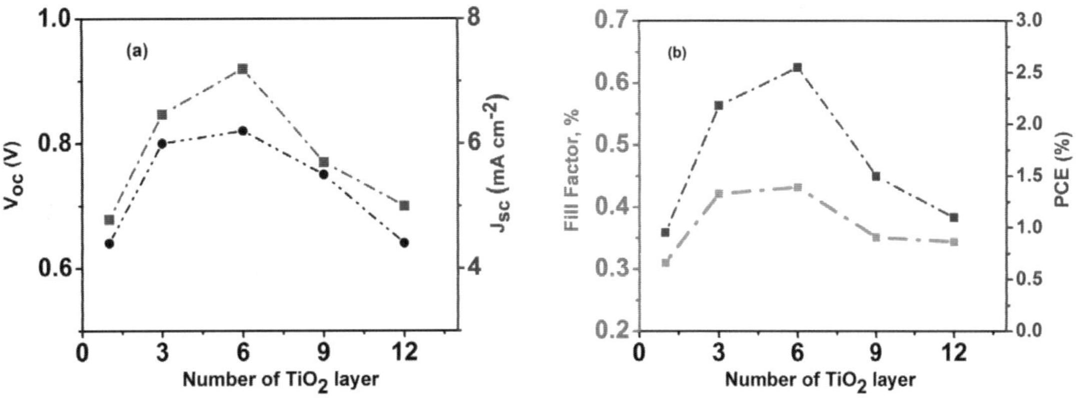

Figure 8. The photovoltaic parameters of the as-fabricated PSCs, (a) V_{oc} and J_{sc}, and (b) FF and PCE.

In order to obtain the TiO_2 thickness, Scanning Electron microscope SEM is used. Figure 9 shows the cross-section SEM image of FTO/TiO_2/ $CH_3NH_3PbI_3$ film that with 6 layers of TiO_2. The thickness of the TiO_2 layer is approximately 108 nm while that of the $CH_3NH_3PbI_3$ perovskite is about 332 nm.

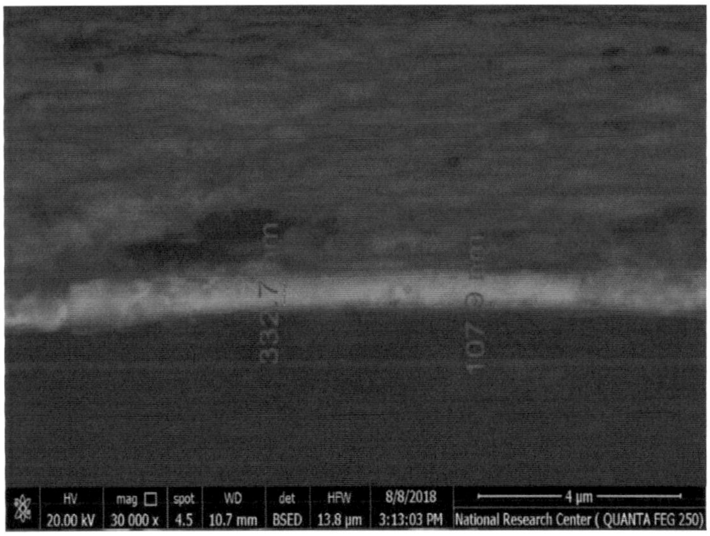

Figure 9: Cross Section SEM image of FTO/TiO_2/ $CH_3NH_3PbI_3$ film.

4. Conclusion

In summary, The Spin coating technique was used to prepare a compact TiO_2 layer with different thicknesses. A perovskite solar cells with structure FTO/TiO_2/ $CH_3NH_3PbI_3$/C were prepared based on different thicknesses of TiO_2 layer. The thicker TiO_2 layer was found to result in absorbing more PbI_2 and therefore relative thicker $CH_3NH_3PbI_3$. We have studied how the TiO_2 layer thickness influences the photovoltaic performance of the perovskite solar cells. The J-V curves revealed that the photovoltaic performance of PSCs could be appropriated with TiO_2 layer of thickness 108 nm.

References

[1] Bi D, Tress W, Dar M I, Gao P, Luo J, Renevier C, Schenk K, Abate A, Giordano F, Correa Baena J-P, Decoppet J-D, Zakeeruddin S M, Nazeeruddin M K, Grätzel M and Hagfeldt A 2016 *Sci. Adv.* **2** e1501170
[2] Huang L, Hu Z, Xu J, Zhang K, Zhang J and Zhu Y 2015 *Sol. Energy Mater. Sol. Cells* **141** 377–82
[3] Back H, Kim J, Kim G, Kyun Kim T, Kang H, Kong J, Ho Lee S and Lee K 2016 *Sol. Energy Mater. Sol. Cells* **144** 309–15
[4] Xing J, Liu X F, Zhang Q, Ha S T, Yuan Y W, Shen C, Sum T C and Xiong Q 2015 Nanolasers *Nano Lett.* **15** 4571–7
[5] Lin K-F, Chang S H, Wang K-H, Cheng H-M, Chiu K Y, Lee K-M, Chen S-H and Wu C-G 2015 *Sol. Energy Mater. Sol. Cells* **141** 309–14
[6] Heo J H, Im S H, Noh J H, Mandal T N, Lim C-S, Chang J A, Lee Y H, Kim H, Sarkar A, Nazeeruddin M K, Grätzel M and Seok S Il 2013 *Nat. Photonics* **7** 486–91
[7] Elshimy H, Abdallah T and Aboshama A 2019 *Int. J. Sci. Eng. Sci.* **3** 15–9
[8] Burschka J, Pellet N, Moon S-J, Humphry-Baker R, Gao P, Nazeeruddin M K and Grätzel M 2013 *Nature* **499** 316–9
[9] Leijtens T, Eperon G E, Pathak S, Abate A, Lee M M and Snaith H J 2013 *Nat. Commun.* **4** 2885
[10] Zhou H, Chen Q, Li G, Luo S, Song T, Duan H-S, Hong Z, You J, Liu Y and Yang Y 2014 *Science* **345** 542 LP – 546
[11] Yang W S, Noh J H, Jeon N J, Kim Y C, Ryu S, Seo J and Seok S Il 2015 *Science.* **348** 1234 LP – 1237
[12] Zhao Y, Nardes A M and Zhu K 2014 *J. Phys. Chem. Lett.* **5** 490–4
[13] Xiao J, Shi J, Li D and Meng Q 2015 *Sci. China Chem.* **58** 221–38
[14] Etgar L, Gao P, Xue Z, Peng Q, Chandiran A K, Liu B, Nazeeruddin M K and Grätzel M 2012 *J. Am. Chem. Soc.* **134** 17396–9
[15] Tu Y, Wu J, Zheng M, Huo J, Zhou P, Lan Z, Lin J and Huang M 2015 *Nanoscale* **7** 20539–46
[16] Yang G, Tao H, Qin P, Ke W and Fang G 2016 *J. Mater. Chem. A* **4** 3970–90
[17] Ye S, Sun W, Li Y, Yan W, Peng H, Bian Z, Liu Z and Huang C 2015 *Nano Lett.* **15** 3723–8
[18] Chen W, Wu Y, Yue Y, Liu J, Zhang W, Yang X, Chen H, Bi E, Ashraful I, Grätzel M and Han L 2015 *Science (80-.).* **350** 944 LP – 948
[19] McMeekin D P, Sadoughi G, Rehman W, Eperon G E, Saliba M, Hörantner M T, Haghighirad A, Sakai N, Korte L, Rech B, Johnston M B, Herz L M and Snaith H J 2016 *Science* **351** 151 LP – 155
[20] El-Brolossy T A, Abdallah S, Abdallah T, Awad H, Mohamed M B, Negm S and Talaat H 2008 *Eur. Phys. J. Spec. Top.* **153** 369–72
[21] Talaat H, Abdallah T, Mohamed M B, Negm S and El-Sayed M A 2009 *Chem. Phys. Lett.* **473** 288–92
[22] Wang H-H, Chen Q, Zhou H, Song L, Louis Z S, Marco N De, Fang Y, Sun P, Song T-B, Chen H and Yang Y 2015 *J. Mater. Chem. A* **3** 9108–15
[23] Li D, Chen Y, Du P, Zhao Z, Zhao H, Ma Y and Sun Z 2015 *RSC Adv.* **5** 88973–8
[24] Badawi A, Al-Hosiny N, Abdallah S and Talaat H 2013 cells *Mater. Sci.* **31** 6–13
[25] Wojciechowski K, Saliba M, Leijtens T, Abate A and Snaith H J 2014 *Energy Environ. Sci.* **7** 1142–7
[26] Adel R, Abdallah T, Moustafa Y M, Al-sabagh A M and Talaat H 2015 *Superlattices Microstruct.* **86** 62–7
[27] Hong S, Han A, Lee E C, Ko K-W, Park J-H, Song H-J, Han M-H and Han C-H 2015 *Curr. Appl. Phys.* **15** 574–9
[28] Hu Q, Wu J, Jiang C, Liu T, Que X, Zhu R and Gong Q 2014 *ACS Nano* **8** 10161–7
[29] Kavan L, Tétreault N, Moehl T and Grätzel M 2014 *J. Phys. Chem. C* **118** 16408–18
[30] Hu H, Dong B, Hu H, Chen F, Kong M, Zhang Q, Luo T, Zhao L, Guo Z, Li J, Xu Z, Wang S, Eder D and Wan L 2016 *ACS Appl. Mater. Interfaces* **8** 17999–8007
[31] Xia J, Masaki N, Jiang K and Yanagida S 2007 *J. Photochem. Photobiol. A Chem.* **188** 120–7
[32] Dai X, Shi C, Zhang Y and Wu N 2015 *J. Semicond.* **36** 74003
[33] Nabil M, Easawi K, Abdallah T, Abdallah S, Elmancy M K, Negm S and Talaat H 2019 *Am. Sci. Res. J. Eng. Technol. Sci.* **53** 27–42
[34] Zhang H, Shi J, Xu X, Zhu L, Luo Y, Li D and Meng Q 2016 *J. Mater. Chem. A* **4** 15383–9
[35] Yella A, Heiniger L-P, Gao P, Nazeeruddin M K and Grätzel M 2014 *Nano Lett.* **14** 2591–6

[36] Bai Y, Dong Q, Shao Y, Deng Y, Wang Q, Shen L, Wang D, Wei W and Huang J 2016 *Nat. Commun.* **7** 12806

[37] Vivo P, Ojanperä A, Smått J-H, Sandén S, Hashmi S G, Kaunisto K, Ihalainen P, Masood M T, Österbacka R, Lund P D and Lemmetyinen H 2017*rg. Electron.* **41** 287–93

[38] Tauc J 1974 Amorphous and Liquid Semiconductor *Amorphous and Liquid Semiconductor* (U.S.: Plenum Press, Springer) pp 159–220

[39] Ghobadi N 2013 *Int. Nano Lett.* **3** 2

[40] Serpone N, Lawless D and Khairutdinov R 1995 *J. Phys. Chem.* **99** 16646–54

[41] Kim H-S, Lee C-R, Im J-H, Lee K-B, Moehl T, Marchioro A, Moon S-J, Humphry-Baker R, Yum J-H, Moser J E, Grätzel M and Park N-G 2012 *Sci. Rep.* **2** 591

[42] Snaith H J, Schmidt-Mende L, Grätzel M and Chiesa M 2006 *Phys. Rev. B* **74** 45306

Size effect of gold nanospheres on the photoacoustic imaging of cancerous cells

H Awad[1], T Abdallah[1], S Negm[2] and H Talaat[1*]

[1] Physics Department, Faculty of Science, Ain Shams University, Abbassia, Cairo, Egypt
[2] Department of Mathematical and Physical Engineering, Faculty of Engineering (Shoubra), Benha University, Cairo, Egypt

* E-mail: hassantalaat@hotmail.com.

Abstract. The size effect of gold nanospheres (AuNS) used as a contrast agent for the photoacoustic imaging (PAI) of breast cancer cell line MDA-MB-231 have been assessed for three different sizes and at four optical densities (ODs). The decrease in the AuNS size at the same OD results in an improved PA signal. The PA signal generally increases linearly with increasing OD, with almost the same rate of increase for the three used particle sizes. It is shown that PEGylation of gold nanoparticles (AuNP) decreases the uptake of the particles by the cancerous cell therefore diminishes the PA images whereas clear images are obtained in the case of unPEGylated AuNPs.

Keywords: Photoacoustic Imaging, Gold nanoparticles, Cancerous cells.

1. Introduction

Photoacoustic imaging (PAI) has recently become of great relevance in biomedical imaging[1-4], and has captured the attention of biomedical engineers, applied physicists, applied mathematicians, and clinical imaging specialists[5]. The technique assembles the biomedical optics and ultrasound which leads to great improvement in imaging [6-8]. The PA technique is based on the photoacoustic effect that was first observed by Alexander Graham Bell in 1880[9]. It uses short- pulsed (or modulated) electromagnetic radiation as probing energy, while detecting ultrasound wave generated due to photon absorption and thermal expansion. There are a lot of advantages of PAI technique over many other imaging techniques, for example it has higher spatial resolution than that of fluorescence and a larger imaging depth, due to the much weaker scattering of ultrasonic signals than light in tissues[10]. Also, the PAI has better tissue contrast than that of ultrasound imaging, since the contrast of the latter is limited by the mechanical properties of biological tissues[11]. Moreover, the absence of ionizing radiation makes this technique safer than Computed Topography and radio nuclide-based imaging techniques[12]. In addition, measuring PA signal at different modulation frequencies one can retrieve individual contribution of scattering and absorption by applying Kubelka –Munk theory [13-16]

Gold nanoparticles (AuNPs) are considered one of the most promising exogenous PA contrast agents [17-20]. This is due to their strong and tunable optical absorption that results from the surface plasmon resonance (SPR) effect. SPR occurs when free charges on the surface of AuNPs oscillate with the electromagnetic field, leading to an optical absorption. Such absorption is several orders of magnitude greater than that of organic dyes[21]. Furthermore, the size and shape of the AuNP determine its resonant frequency, and absorption cross section which allows for fine tunability[22-24]. In this study, we have carried out an investigation of the effect of the size of AuNPs as a contrast agent on the PA imaging. The used particles were gold nanospheres(AuNS) of three different sizes ~12 nm, 66 nm, and 90 nm, that were synthesized by seed-mediated growth technique (as the larger sizes produced by citrate reduction were not stable or spherical). The AuNS were characterized by UV-Vis absorption, transmission electron microscope (TEM), Dynamic light scattering (DLS), and Zeta potential measurements. The AuNPs are then applied to breast cancerous cells of type MDA-MB-231 at different optical densities (ODs) to assess the effect of AuNP size on the PA signal.

2. Experimental Setup

2.1. Synthesis and characterization of gold nanospheres with different sizes

AuNS were synthesized using the seed-mediated growth technique as described in[25]. A 100 ml volume of 0.27 mM $HAuCl_4$ in deionized water was brought to a boil. A 5.28 ml volume of 34 mM sodium citrate was added to the solution under vigorous stirring and allowed to stirr for several minutes. Then the solution temperature was allowed to return to room temperature. The synthesized nanospheres were of size ~ 12 nm. Larger nanospheres were synthesized using 7.5 ml of 25 mM $HAuCl_4$ mixed with 15.61 ml of 0.2 M hydroxylamine hydrocloride $NH_2OH.HCl$ and 750 ml of deionized water. A 50ml and 12.5 ml of the as-prepared seed solution were added to the solution under vigorous stirring. After nanoparticles were allowed to sit overnight, they were characterized by UV–Vis spectroscopy (Ocean Optics), to determine their optical absorption properties. Zetapotential (Zeta sizer Nano Series, Malvern) was used to determine the surface charge. The particle size was determined using dynamic light scattering DLS, and transmission electron microscope TEM (JEOL 100 CX-II).

A solution of 1mg/ml methoxy polyethylene glycol thiol (mPEG-SH) of 5K molecular weight was added to same volume of a solution of citrate-capped Au nanoparticles in order to increase the particle stability and biocompatibility[26]. The solution was shaken at lowest speed overnight allowing citrate ligands to exchange with mPEG-SH. The excess mPEG-SH was removed via centrifugation at 7000 rpm for 10 minutes. The resulting colloidal solutions were stable (for several months) as there is no change in the UV-Vis absorption spectrum measured after 3 months.

The samples were inspected using a JEOL 100 CX-II TEM operating at 100 kV. The TEM imaging of gold nanoparticles were carried out by placing the nanoparticles on carbon-coated copper grid and dried overnight in air. All of the micrographs were recorded on a CCD camera, and the images were analyzed using Image J software to get the particle size and the standard deviation. Hydrodynamic size for the prepared particle was measured by Malverm Zeta sizer using the default non-invasive back scattering (NIBS) technique at 25 ^{o}C. The scattered light is measured using a detector at an angle of 173° for the particles sizes used (according to the instructions).

2.2. Cell culture

Breast cancer cell line, MDA-MB-231, used in this work to assess the effect of the size of gold nanospheres as a contrast agent on PA imaging, were cultured in Dulbecco's modified eagle medium: Nutrient Mixture (DMEM/F12) supplemented with 10% fetal bovine serum, 2% pencilline/streptomycin sulphate and incubated at 37 ^{o}C with 5% CO_2 at 95% relative humidity. Media were exchanged every 2–3 days and the cells were passaged with trypsin when 90% confluent. Cells are then counted using a hemocytometer and reseeded in 6-well cell culture plates of about $5x10^5$ cell per well, and then allowed to adhere overnight prior to the addition of nanoparticles. For each of the three different sizes, four different OD's of 0.5, 1, 3, and 4, were prepared in phenol red-free DMEM/F12 cell media Optical Density (the optical attenuation per cm of solution OD). The particles were then incubated with the cells for 22 hours. After incubation; the media was aspirated, cells were then washed with phosphate buffered saline (PBS) 3 times to get rid of any unattached particles to the cells. In order to fix the cells in the plates; 2ml of 10% formalin were added for 10 minutes. The loaded cells are then covered with PBS and kept in the fridge for the experiment. For combined Ultrasound (US)-PA imaging, the loaded cells passaged with trypsin and the trypsin is then quenched with media. The cells are then centrifuged and fixed with 10% formaline, centrifuged and resuspended in cold PBS. A Tissue mimicking phantom with loaded cell inclusions was prepared for photoacoustic imaging. The cell suspension (20 μL) was mixed with gelatin solution (6% by weight after mixing) and placed on the base layer of the gelatin phantom.

IOP Conf. Series: Materials Science and Engineering **762** (2020) 012004 doi:10.1088/1757-899X/762/1/012004

2.3. PA measurements

The PA images for the tissue-mimicking phantom were obtained using a combined ultrasound (US)-PA transducer. US and PA signals were captured using a 40 MHz array transducer with a Vevo 2100 ultrasound micro imaging system (VisualSonics, Inc.) as is schematically presented in Figure 1a. A tunable OPO laser beam (Premiscan, GWU, Inc.) pumped by a pulsed Nd:YAG laser (Quanta-Ray, Spectra Physics, Inc.) was delivered through an optical fiber bundle (Ceramoptec, Inc.). The fiber bundle was combined with the ultrasound transducer and aligned such that the transmitted light would focus at the focal point of the array transducer. The trigger signal from the laser system was synchronized with the Vevo 2100 imaging system to capture the photoacoustic signal when the pulsed laser irradiated. In general, combined US and PA images when overlayed would indicate the locations of aggregated AUNS on the malignant cells on the background of the US topographic image. The experimental set up used for measuring only the PA signal is shown in Figure 1b. In this experiment, the excitation is generated by a Q-switched, pulsed Nd: YAG laser (Quanta-Ray PRO-290, Spectra-Physics Lasers, Mountain View, CA, USA). The laser output was at a wavelength of 532 nm, with a pulse-width of 5-7 ns and a pulse repetition rate of 10 Hz. The laser beam is split to two beams, the first beam is measured with a powermeter, and the second beam is focused on the sample using a plano-convex lens. PBS is used as an acoustic coupling medium to reduce the acoustic impedance between the sample and the ultrasound receiver. The induced PA pressure wave is measured by a 1 mm needle hydrophone with 4 dB and bandwidth of 0.2 - 15 MHz (Precision Acoustics LTD, Dorchester, UK). The PA signal acquired by the hydrophone is amplified by an ultrasound receiver amplifier and then digitized using an oscilloscope. The laser fluence was about 5 mJ/cm^2, and the PA signal was measured at a 100 different points in each well.

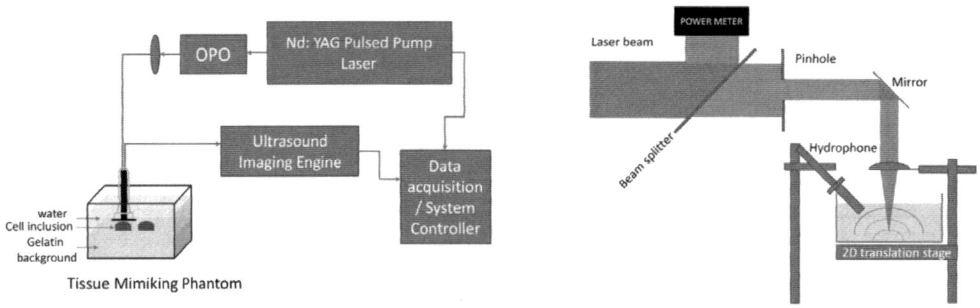

(a) (b)

Figure 1 block diagram of (a) combined US-PA imaging system and the gelatin tissue mimicking phantom with gold nanoparticle labeled cell inclusions , and (b) the set up used for measuring PA signal of gold nanoparticles labeled cells in 6 well plate

3. Results and Discussion

Three different sizes of AuNSs were synthesized and the UV-Vis spectra were obtained and the sizes were determined both by TEM and DLS. The UV-Vis spectra for gold nanospheres of different sizes are shown in Figure 2a. As the particles size increases, the absorption peak due to SPR is shifted to longer wavelength, the maximum wavelength λ_m for the 12 nm, 66 nm and 90 nm nanoparticles are observed at 522, 538 and 556nm respectively. Furthermore, the peak width at the base is increased with increasing size which can be understood in terms of the fact that higher oscillation modes (quadrupole, octopole, absorption, and scattering) also affect the extinction cross section as the size increase[27,28]. The UV-Vis absorption spectra for citrate coated 12 nm Au spheres before and after PEGylation are shown in Figure 2b. It is observed that the SPR peak is red shifted about 3 nm after PEGylation, as indicated in the inset of Figure 2b which clearly indicates the PEG surface modification.

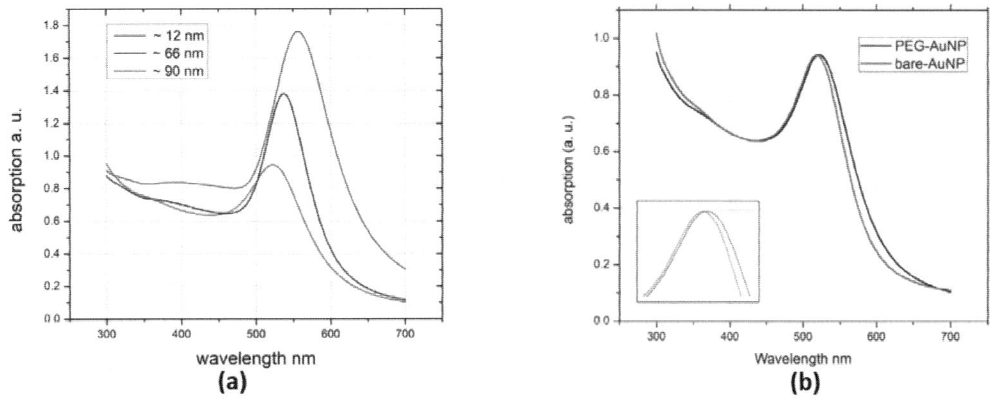

Figure 2 (a) UV-Vis Absorption spectra of Au Nanoparticles of the three different sizes, (b) UV-Vis Absorption spectra of the 12 nm Au Nanoparticles before and after PEGylation, a clear red shift of about 3 nm after PEGylation is shown in the inset.

The TEM image for two particle sizes (12nm, 66nm) are shown in Figures 3a and 3b respectively. The size distribution histogram for the smaller size is shown in Figure 3c having an average size of 11.8± 1.6 nm and for the larger size in Figure 3d having an average size of 65.92± 6.11 nm.

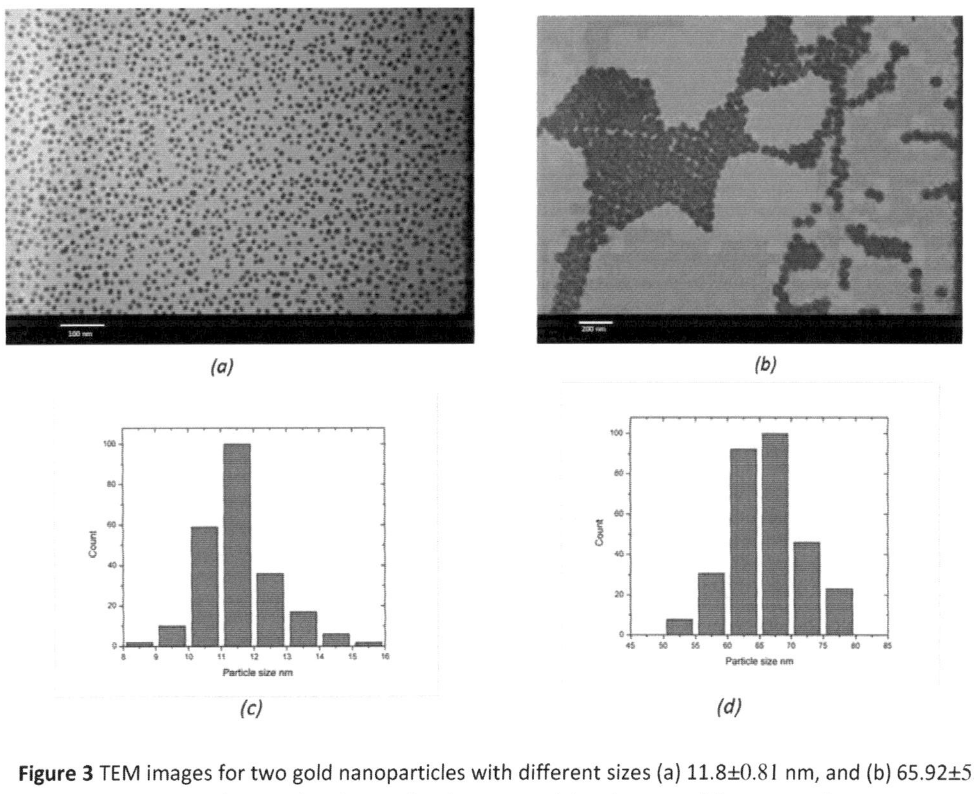

Figure 3 TEM images for two gold nanoparticles with different sizes (a) 11.8±0.81 nm, and (b) 65.92±5.32 nm, the size distribution for the ~ 12nm (c) and ~66 nm (d) respectively.

The hydrodynamic size for the AuNS was averaged over three measurements for each sample giving values for the particle sizes of 15.87 nm, 70.35 nm, and 98.06 nm with a poly dispersity index of about 0.08 as indicated in Figure 4 by a, b, and c respectively. Those values of hydrodynamic sizes are slightly different than the values obtained using TEM due to the presence of surfactant and the interaction with solvent molecules.

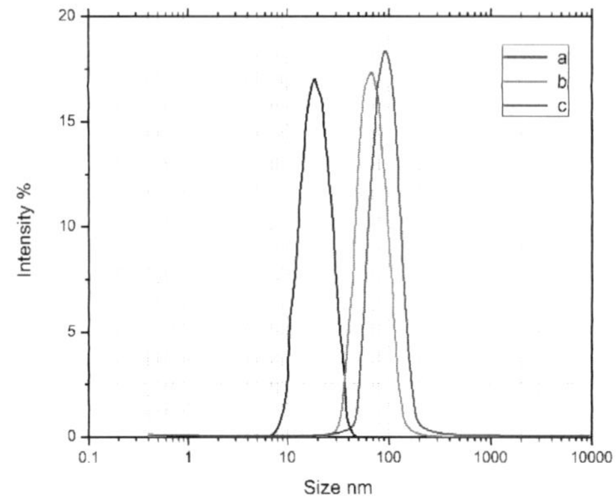

Figure 4 Hydrodynamic size distribution for AuNS of sizes (a) 16.12±7.5, (b) 70.53±21.48, and (c)98.06±28.03nm

The Zeta potential measurements were carried out in order to confirm the surface modification of AuNS for citrate stabilized particles after and before PEGylation. The citrate coated AuNPs have a negative value of -34.3 mV as shown in Figure 5a. This is due to the attached citrate. After PEGylation, the Zeta potential is reduced to -9.35 mV as shown in b of Figure 5. AuNS of large size, obtained by addition of $NH_2OH.HCl$ as a reducing agent, have a positive surface charge of about +19.8 mV their plot is represented in Figure 5c.

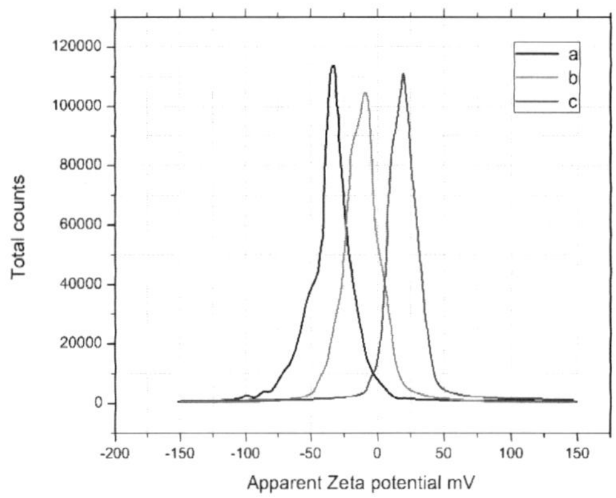

Figure 5 Zeta potential of (a) citrate stabilized AuNS, (b) PEGtlated AuNS, and (c) AuNS sabilized with $NH_2OH.HCl$

The US and PA images for the tissue mimicking phantom with cell inclusions loaded with citrate stabilized AuNPs are shown in Figure 6a on the left and right sides, respectively. Figure 6b shows the images for cells with PEGylated AuNPs. There is no difference between the US images for cells incubated with unPEGylated or PEGylated AuNS as the US image provides only a structural information about the tissue. While the PA image shows a very bright image for cells loaded with citrate stabilized AuNS, on the other hand there is almost no image in case of PEGylated AuNS. Such large increase in the PA signal may be due to the higher aggregation of citrate capped nanospheres on the cells than that of the PEGylated particles. This was also confirmed by the bright field microscope image taken for the cells loaded with the citrate shown in Figure 6c and PEG capped particles shown in Figure 6d. The two images were taken after incubation of the particles for the same period of time in media and at the same OD. Figure 6c shows a highly accumulated citrate capped nanoparticles on the cells, whereas Figure 6d shows PEGylated particles of the same size at a much lower accumulation. As mentioned in the experimental

part, AuNPs were incubated in culture medium that contains 10% Fetal Bovine Serum which results in binding of serum protein to the surface of nanoparticles. The binding of serum proteins results in surface modification and increasing intracellular uptake by receptor-mediated endocytosis[29, 30]. Cui et al [31]studied the interaction of different AuNPs with serum proteins. Their results shows that citrate capped NPs are capable of adsorbing bovine serum albumin to form a 6-8 nm thickness protein corona. After modification with high densities of hydrophilic PEG, AuNPs have a high capacity to resist the adsorption of proteins. Therefore, these AuNPs circumvent cellular uptake. It should be mentioned that while the combined US-PA transducer scans the whole sample at once, the PA signal allows for taking measurements at different spots in the sample giving a large number of data to be averaged. The PA signal is measured for the cells while they are fixed in the 6 well cell culture plate , the signal is measured at 100 different points for each well to increase the precision. The PA signal for MDA-MB-231 Breast cancer cells loaded with 12 nm citrate stabilized AuNS, was first obtained at OD 0.5, 1, 3, and 4 and plotted as shown in Figure 6e. The PA signal for cells loaded with the same size but PEGylated spheres at the same OD was also plotted for comparison. It is observed that the citrate capped (unPEGylated) AuNs gives higher PA signal that is about 10 times more than the PEGylated particles for all ODs. This result explain the reduced cell uptake of the nanoparticles by MDA MB 231 breast cancer cells. Therefore it was decided to use unPEGylated particles in the rest of our study.

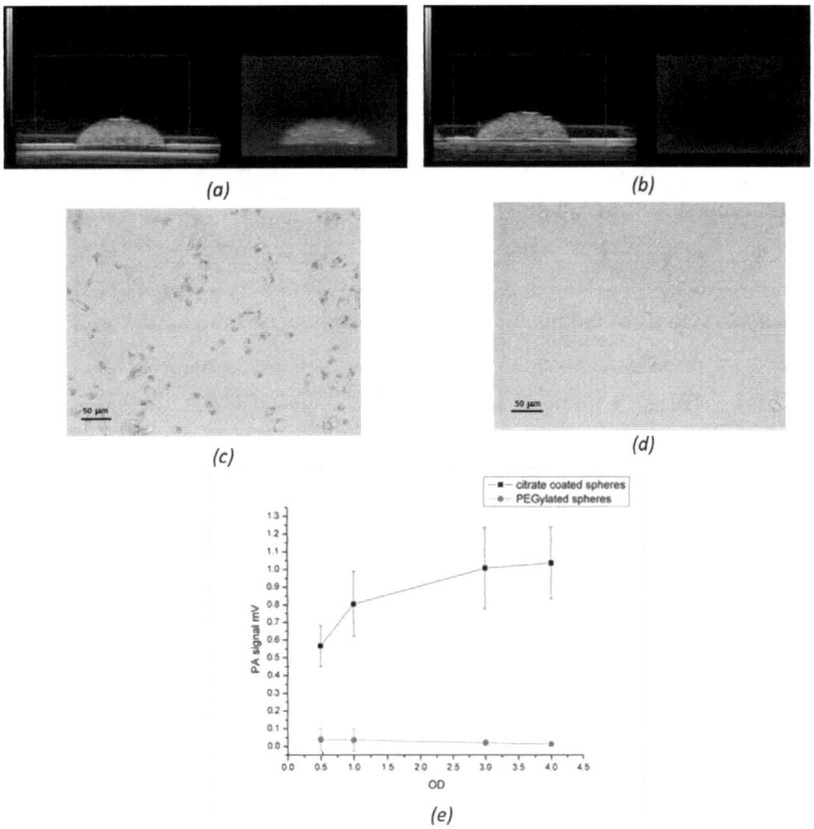

Figure 6 US (left) and PA (right) images for cells loaded with (a) 12 nm citrate stabilized and (b) PEGylated AuNPs, the US image in the left side, and the PA image in the right side, bright field microscope images for gold nanoparticles (c) capped with citrate, and (d) PEGylated AuNPs. (e) Variation of the PA signal with OD for cells

loaded with AuNPs capped with citrate, and PEGylated gold nanoparticle as indicated there is no observed PA signal for PEGylated AuNS at any OD.

Furthermore, dark field microscope images were taken for the cells loaded with two different concentrations of NPs but of the same size. Cells incubated with higher optical densities (OD 3) (Figure 7a) shows a higher light scattering than cells incubated with particles of smaller optical densities (OD 1) (Figure 7b). The variations of the PA signal with OD for three different sizes of AuNPs designated as small ~12 nm, medium ~66 nm, and large ~90 nm are shown in Figure 8. For each particle size, there is an increase in the PA signal with the increase in the OD. Furthermore, for the same amount of gold on the cells (same OD), it is observed that the PA signal increase as the size of the AuNPs is decreased giving the highest PA signal for the smallest size 12 nm. This is understood in terms of the increase of surface area for the smaller particle size. At the same OD 1, the surface area for particle sizes of 10nm, 65 nm, and 90 nm are given by 17.9 cm^2, 2.9 cm^2, and 1.98 cm^2 respectively[32]. This surface area variation with the particle size agrees with the variation in the PA signal obtained within experimental error. The aggregation of AuNPs causes the conduction electrons near each particle surface to become delocalized and be shared amongst neighboring particles. This leads to the SPR to shift to lower energy causing the absorption and scattering peaks to red shift[33]. Also, the aggregation may help in formation of more hotspots for the smaller nanoparticles. With decreasing the particle size, the incident photons may undergo more diffuse scattering within the sample which results in an increase in absorption and consequently the PA signal.

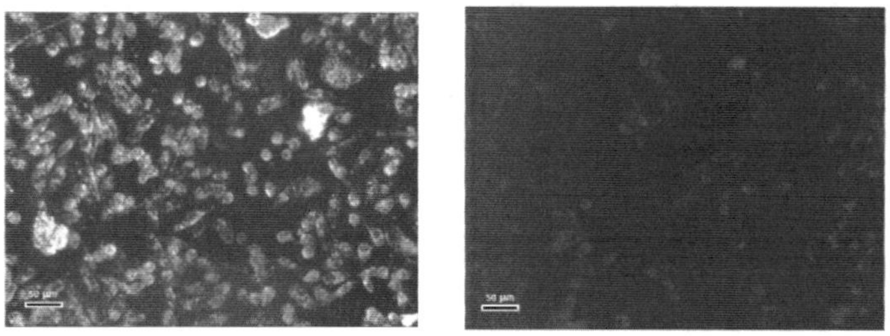

(a) *(b)*

Figure 7 Dark field microscope image for cells after incubation with AuNPs with OD3 *(a)* and OD1 *(b)*.

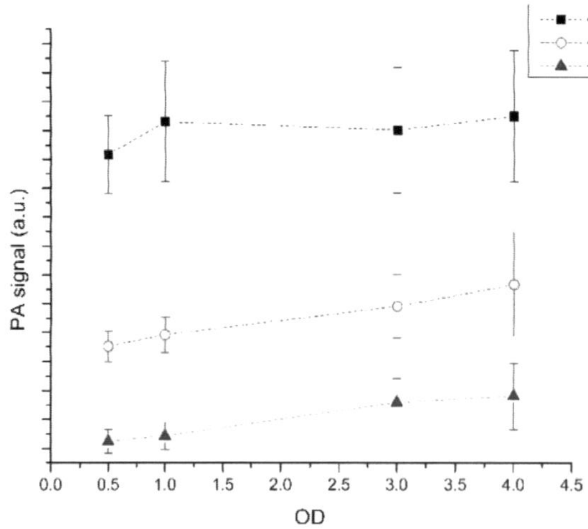

Figure 8 PA signal for cells loaded with gold nanoparticles with different ODs for three different particle sizes as indicated.

4. Conclusions

Using AuNPs as a contrast agent for PA imaging of MDA-MB 231 breast cancer cells, it is found that the PEGylation of nanoparticles reduce cell uptake than the citrate capped particles. This is because the PEGylated AuNPs have a higher capacity to resist the adsorption of proteins in the cell culture media which reduce the cellular uptake. Therefore, In order to employ PEGylated nanoparticles effectively, the nanoparticles have to be conjugated to a peptide to penetrate the cell to give a larger PA signal. The PA signal from AuNPs used as contrast agent, increases linearly with increasing the OD of NPs incubated with the cells. The PA signal obtained for cells loaded with AuNPs is significantly larger for smaller particles than larger size particles for the same OD as a result of the increase in their surface area. So, it is recommended to apply small AuNPs (~12 nm) as a PA contrast agent to obtain a more bright image with less amount of gold to avoid cell toxicity. The possible impact of the surfactant surface charge on the cell uptake of NPs and hence its effect on the PA signal is the subject of future study.

Acknowledgment

The authors greatly appreciate Prof. Stanislav Emelianov, Georgia Institute of Technology for hosting H. A. at his laboratory. The authors also thank the generous support of the Egyptian STDF grant ID 3746.

References

[1] Lutzweiler C, Razansky D 2013 *Sensors* **13**(6) 7345-7384
[2] Wang L V, Hu S 2012 *Science* **335**(6075) 1458-1462
[3] Mallidi S, Luke G P, Emelianov S 2011 *Trends in biotechnology* **29**(5) 213-221
[4] Beard P 2011 *Interface focus* **1**(4) 602-631
[5] Manohar S, Razansky D 2016 *Advances in optics and photonics* **8**(4) 586-617
[6] Ntziachristos V 2010 *Nature methods* **7**(8) 603-614
[7] Niederhauser J J et al. 2005 *IEEE transactions on medical imaging* **24**(4) 436-440
[8] Cook J R, Bouchard R R, Emelianov S Y 2011 *Biomedical optics express* **2**(11) 3193-3206
[9] Bell A G 1880 *Journal of the Society of Telegraph Engineers* **9**(34) 404-426
[10] Pei Y et al. 2014 *Scientific reports* **4**
[11] Cai X et al. 2011 *Journal of biomedical optics* **16**(4) 046017-046017
[12] Li W, Chen X 2015 *Nanomedicine* **10**(2) 299-320
[13] Lamastra F et al. 2018 *International Journal of Thermophysics* **39**(9) 110
[14] Voti R L et al. 2015 *International Journal of Thermophysics* **36**(5-6) 1342-1348
[15] Leahu G et al. 2017 *Scientific reports.* **7**(1) 2833
[16] Lamastra F et al. 2017 *Nanotechnology* **28**(37) 375704
[17] Copland J A et al. 2004 *Molecular Imaging & Biology* **6**(5) 341-349
[18] Zhang Q et al. 2009 *Nanotechnology* **20**(39) 395102
[19] Agarwal A et al. 2007 *Journal of applied physics* **102**(6) 064701
[20] Abdallah T et al. 2012 *J. Raman Spectrscopy* **43**(12) 1924-1930
[21] Jain P K et al. 2006 *J. Phys. Chem. B* **110**(14) 7238-7248
[22] Kelly K L et al. 2003, ACS Publications. **107**(3) 668-677
[23] Lee K S, El-Sayed M A 2006 *The Journal of Physical Chemistry B* **110** (39) 19220-19225
[24] El-Brolossy T A et al. 2008 *European Physcis Journal: Special Topics* **163**(1) 361-364
[25] Bayer C L, Kelvekar J Emelianov S Y 2013 *Nanotechnology* **24**(46) 465101
[26] Rahme K et al 2013 *Rsc Advances* **3**(17) 6085-6094
[27] Link S, El-Sayed M A 1999 *The Journal of Physical Chemistry B* **103**(21) 4212-4217
[28] Dreaden E C et al. 2011 *Applied Physics Letters* **98**(18) 183115
[29] Chithrani B D, Ghazani A A Chan W C 2006 *Nano letters* **6**(4) 662-668
[30] Chithrani B D, Chan W C 2007 *Nano letters* **7**(6) 1542-1550
[31] Cui M et al. 2014 *Nano Research* **7**(3) 345-352
[32] Solutions b 2018 *gold nanoparticles*:

http://www.bbisolutions.com/en/technical-support/gold/nanoparticles/molar-concentration-of-gold-nanoparticles-colloid.html.
[33] Fratoddi I et al 2017 *The Journal of Physical Chemistry C* **121**(33) 18110-1811

Preparation of Silver Nanoparticles Dispersed in Almond Oil Using Laser Ablation Technique

Romany Onsi [1], Khaled Easawi [1], Said Abdallah[1], Sohair Negm[1] and Hassan Talaat[2]

[1] Laser physics & technology unit, Faculty of Engineering (Shoubra), Benha University, Cairo, Egypt
[2] Physics Department, Faculty of Science, Ain Shams University, Abbassia Cairo, Egypt

* E-mail: drnegm@hotmail.com

Abstract. Nanoparticle production by pulsed laser ablation (PLA) is a process that can generate pure nanoparticles (NPs) straight from a varied range of bulk substances and compounds. Silver nanoparticles (Ag NPs) are probably one of the most attractive noble metal nanostructures because of their unique and interesting physical and chemical properties. In this study, laser ablation of pure silver plate immersed in almond oil was carried out for Ag NPs production. Nd: Yag laser of wavelength 1064 nm, was used for the ablation of the Ag plate at different laser energies and ablation times. The almond oil permitted the formation of Ag NPs with a stable and homogeneity particle diameter in a reasonable time. The size distribution of the NPs was examined by High-Resolution Transmission Electron Microscopy (HRTEM). The particle sizes of the produced Ag NPs at laser energy 200 mJ in the solution at 10,15 and 20-min ablation times were 4.82, 3.11 and 1.82 nm respectively. The particle sizes of Ag NPs produced at different laser energies 150,200 and 250 mJ and at ablation time 10 min inside the solution were 5.39,4.82 and 1.92 nm respectively. The absorption peaks of the produced nanoparticles have been characterized using a UV-Vis spectrophotometer.

Keyword: Laser ablation, metal nanoparticle, almond oil

1. Introduction

Nanoparticles (NPs) formed by conventional chemical synthesis methods often include undesirable residues from the reactants. These residues can be toxic and unfavorable for applications. Production of NPs by pulsed laser ablation in liquid medium (PLAL) has received much attention as a novel NPs production method. The most important advantages of NPs prepared by the PLAL method are its high stability, the relative simplicity of the procedure and the absence of chemical reagents in the final preparation [1][2][3]. Due to their fascinating optical, catalytic, electronic and magnetic properties, which are particularly different from bulk samples, noble-metal NPs have seen spectacular growth in recent years. Silver NPs are probably the most attractive nanostructures for noble metals[4].

In this study, we prepared silver NPs in almond oil of different sizes. The main problem of the production of Ag NPs is the agglomeration of Ag nanoparticles in liquid due to thermodynamics, so recently some fatty acids or vegetable oils were used as a stabilizer[5] Hence, we have used almond oil in this study as a stabilizing agent in the preparation of Ag nanoparticles. Sweet almond oil has been described as unsaturated oil, with oleic acid (O, C18:1) being the main fatty acid (65%) with β – sitosterol as the most representative sterol and α – tocopherol as the major tocopherol[6][7][8]. We were able to control the size of produced Ag nanoparticles in almond oil by changing the laser parameters to obtain Ag NPs of the size required for a cosmetic application[9][10]. The size distribution of the NPs was determined by High-Resolution Transmission Electron Microscopy (HRTEM).

2. Experimental setup

The experimental setup is shown schematically in Fig. 1. The beam from a Pulsed Q-Switched Nd: YAG (Quanta - Ray) laser of wavelength 1064 nm, pulse duration 8 ns and repetition rate 10 Hz was focused by a lens of focal length (f = 20 cm) on the surface of the silver target (purity: 99.99 % from Sigma Aldrich). The target was of a thickness (0.5 mm) and dimensions (2.5*2.5 cm^2) placed at the bottom of a glass vessel containing (20 ml) of almond oil (from FLUKA). The metal plate was washed first using the ultrasonic bath for 30 min and cleaned with ethanol and distilled water to get rid of organic contamination and then fixed at 18 mm below the almond oil surface.

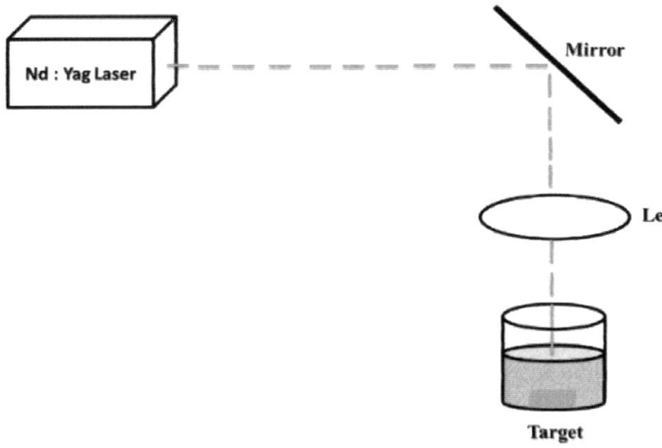

Fig 1. Schematic diagram of the experimental setup

The energy of the incident laser pulse was measured using (COHERENT- LASERMATE/D) power meter. The effective spot size of the laser beams on the surface of the target was 7 mm without the lens and is calculated to be 2mm on the surface of the target_for wavelength (1064 nm). Ablation was carried out at room temperature and the formation of nanoparticles in the almond oil could be confirmed by the change of its color during ablation. During the ablation procedure, the solution was magnetically stirred to disperse the generated NPs[11]. The absorption spectra of the Ag nanoparticles in the almond oil were measured immediately after the ablation process using a UV-Vis double-beam spectrometer type (JASCO V-670). High-Resolution Transmission Electron Microscope (HRTEM)(JEM-2100) was used to determine the shape and the size distribution of the nanoparticles[12].

3. Results and Discussion

Through the laser ablation of silver plate in the almond oil, the almond oil firstly appeared colorless but after a small period of time, it began to change to yellow, then finally to brown. The darker color for solutions suggests a higher concentration of Ag NPs and this was confirmed by a UV-Vis absorption spectrum.

The FT-IR spectra of pure almond oil and of silver NPs in almond oil are shown in Fig.2 where the peaks corresponding to bands of olefinic double bonds (3002 cm-1), methylene group (2922, 2851, 1464 and 722 cm-1), methyl group (1464 and 1377 cm-1) and triglyceride ester group (1743, 1237, 1159 and 1097 cm-1)[13]. Actually, the peaks are sharper in the case of Ag NPs in almond oil due to the plasmonic effect of the nanoparticles[14][15]. We can say that the almond oil component can be used as a capping agent of Ag NPs.

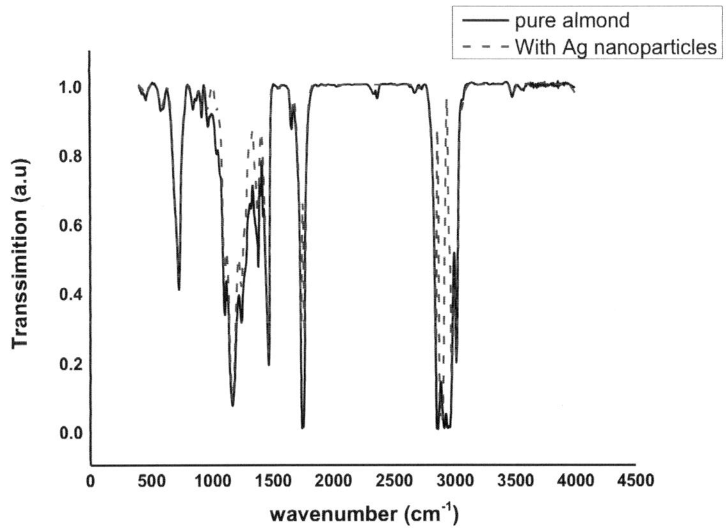

Figure 2. The FT-IR spectrum of pure almond oil and Ag nanoparticles in almond oil

3.1 Effect of ablation time

The optical absorbance spectra of silver colloids prepared by laser ablation in almond oil solution at different ablation times ranged from (5 to 20) min are shown in fig. 3.

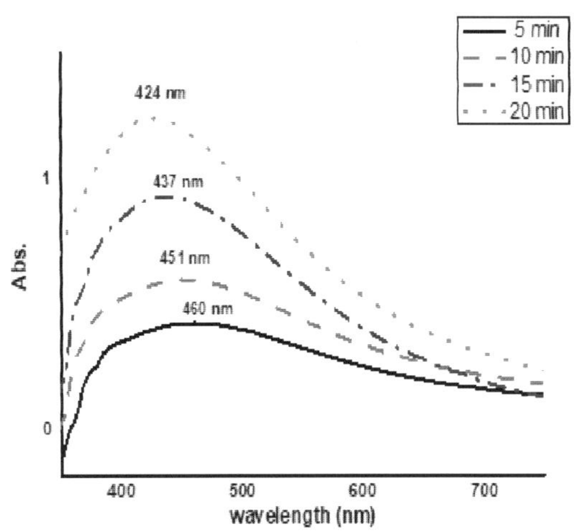

Figure 3. UV-Vis results show the intensity peaks having a blue shift with Increasing ablation time at laser energy 200 mJ

It is noticeable in fig.3 that, increasing the time of ablation from 5 to 20 minutes cause the absorption peaks to increase and shift to lower wavelength (blue shift). The absorption peaks arise from surface plasma resonance of silver nanoparticles. The increase in the peak intensity of the spectra is due to the increase in the concentration and volume fraction of the Ag nanoparticles inside the liquid. Conferring to Mie theory, the blue shift in spectra is due to the decreases in particle size[16][17]. This decrease in particle size was due to the interaction between produced Ag NPs and laser light for longer ablation times[18]. On the path of incident laser light, many Ag NPs were produced. Due to the interaction of the laser light and produced Ag NPs, large particles break down and get smaller. Fragmentation quality improves with increasing removal time; so, the obtained particles at longer ablation times are smaller and significant aggregation is not seen, which is expressed by the narrower particle distribution.

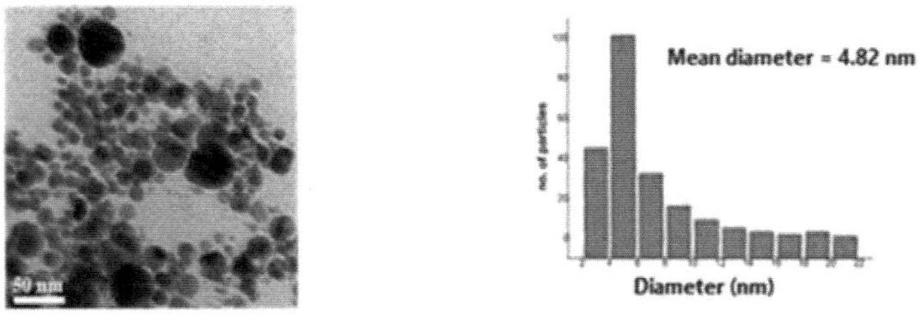

a) TEM micrograph and histogram for silver nanoparticle with time (10 min) (average diameter = 4.82 nm)

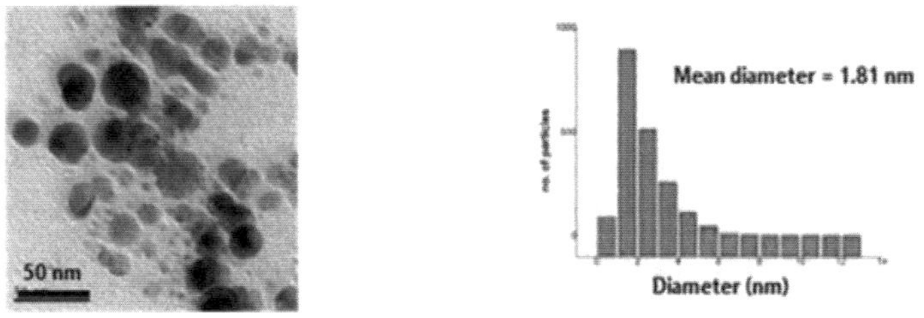

b) TEM micrograph and histogram for silver nanoparticle with time (20 min) (average diameter = 1.81 nm)

Figure 4. TEM micrograph and histogram for Silver nanoparticles in almond oil with times (10 and 20 min)

The TEM micrograph and size distribution of silver nanoparticles produced by laser for ablation time 10 and 20 min in almond oil solution at laser energy 200 mJ are shown in fig. 4. The corresponding average size of the produced NPs with 10 min ablation time was 4.82 nm and with increasing the ablation time to 20 min the size decreased to 1.81 nm.

3.2 Effect of the energy of the laser beam (150 -250 mj)

The absorption spectra of Ag NPs in almond oil at different laser energies ranged from (150 to 250) are shown in fig. 5. It is observed that the absorption peak position is slightly shifted towards the shorter wavelength with an increase in laser energy (blue shift). This finding suggests a reduction in particle size of the Ag NPs prepared with an increase in laser energy. In fact, higher laser energy leads to greater kinetic energy which increases collision between initially formed large nanoparticles in the condensation process which further reduces the size[19]. It is observed that the absorption peak intensity increases with the increase of laser energy. This attributed to the increase in the concentration and volume fraction of the Ag NPs produced in the almond oil.

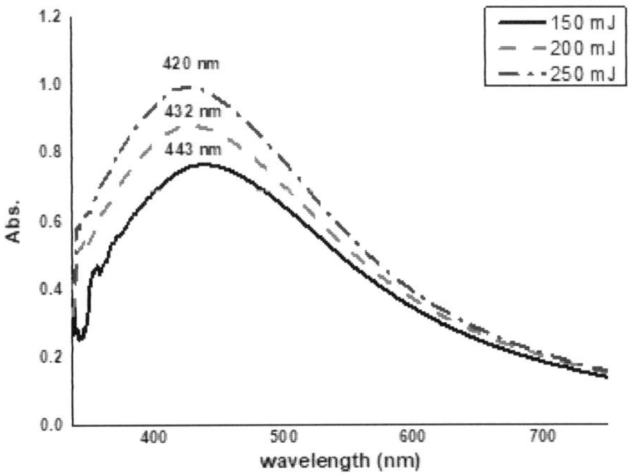

Figure 5. UV-Vis results show the intensity peaks having a blue shift with increasing laser energy from 150 to 250 mJ **at 10 min ablation time**

The TEM micrographs and size distribution of Ag NPs produced by laser ablation for energy 150 and 250 mj at ablation time 10 min in almond oil solution. The average size decreases from 5.39 nm to 1.92 nm with increasing laser energy from 150 to 250 Mj are shown in fig 6.

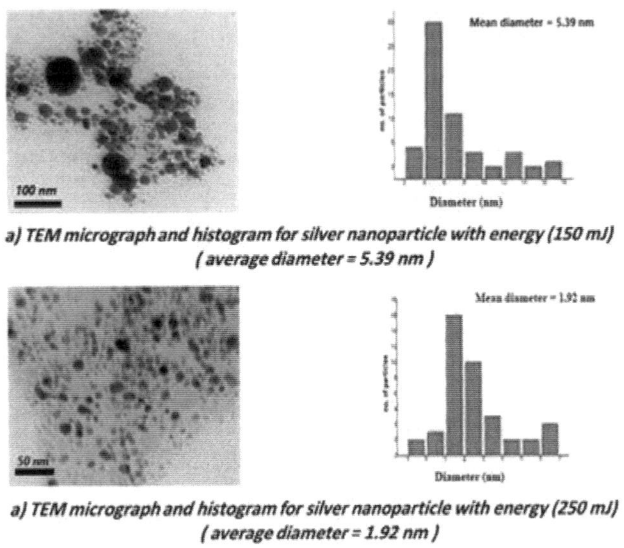

Figure 6. TEM micrograph and histogram for Silver nanoparticles in almond oil generated with energy (150 and 250 mJ)

3.3 Effect of aging

The absorption spectra of fresh (Ag/almond oil) and (Ag/almond oil) measured after 2 weeks and 4 weeks are shown in fig7. All samples display a little shift in peak position and a small increase in the full width of the half maximum (FWHM). This indicates minute aggregation of the ablated Ag NPs in almond oil.

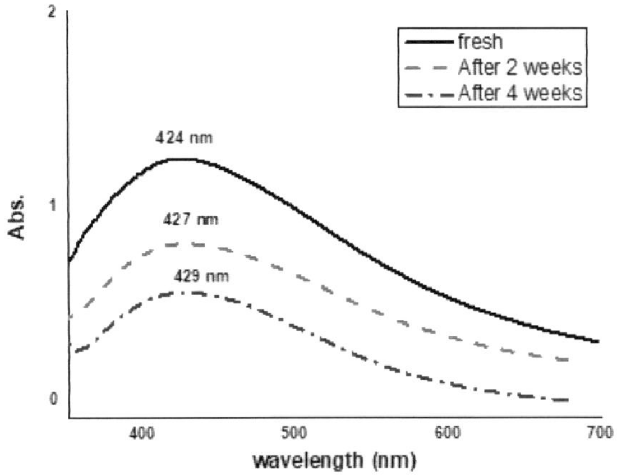

Figure 7 .UV-VIS for fresh and after several weeks

This observation shows the reasonable stability of the samples and also confirms the ability of almond oil as a stabilizer and size controller. These effects are due to the rivalry between the rapid formation of initial silver nanoparticles and the consequent growth of the particle due to the almond oil molecules capping the particle[20]. We can say that at the beginning of the ablation of the silver plate, A thick cloud of silver atoms over the ablation spot is produced. Silver atoms interact much stronger than silver atoms and almond oil molecules[21]. Therefore, a very slow agglomeration of the silver atoms occurs.

4. Conclusion

Laser ablation of a silver plate in almond oil was employed to prepare size-controlled silver nanoparticles using, fixed laser energy with increasing ablation time period and increasing laser energy for a fixed time period. The size of the ablated Ag NPs decrease in both cases, furthermore, the almond oil proved to be a good stabilizer for Ag NPs. Such as a simplified method for generating Ag nanoparticles in almond oil is relevant importance for the cosmetic industry.

References

[1] Nikov R G, Nikolov A S, Nedyalkov N N, Dimitrov I G, Atanasov P A and Alexandrov M T 2012 Applied Surface Science Stability of contamination-free gold and silver nanoparticles produced by nanosecond laser ablation of solid targets in water *Appl. Surf. Sci.* **258** 9318–22

[2] Darwish A M, Eisa W H, Shabaka A A and Talaat M H 2015 Synthesis of Nano-Cadmium Sulfide by Pulsed Laser Ablation in Liquid Environment *Spectrosc. Lett.* **48** 638–45

[3] Darwish A M, Eisa W H, Shabaka A A and Talaat M H 2016 Investigation of factors affecting the synthesis of nano-cadmium sulfide by pulsed laser ablation in liquid environment *Spectrochim. Acta - Part A Mol. Biomol. Spectrosc.* **153** 315–20

[4] Solati E and Mashayekh M 2013 Effects of laser pulse wavelength and laser fluence on the characteristics of silver nanoparticle generated by laser ablation 689–94

[5] Press D 2011 Preparation of silver nanoparticles in virgin coconut oil using laser ablation 71–5

[6] Cherif A, Belkacemi K, Kallel H, Angers P and Arul J 2009 Phytosterols , unsaturated fatty acid composition and accumulation in the almond kernel during harvesting period : Importance for development regulation *C. R. Biol.* **332** 1069–77

[7] Content O, Composition F A and Seeds D A 1988 Oil Content and Fatty Acid Composition 695–7

[8] Beltra A, Pe S E M and Fame Á G C Á 2009 Classification of Four Almond Cultivars Using Oil Degradation Parameters Based on FTIR and GC Data 51–8

[9] Imam H, Elsayed K, Ahmed M A and Ramdan R 2012 Effect of Experimental Parameters on the Fabrication of Gold Nanoparticles via Laser Ablation **2012** 73–84

[10] Physics A 2004 Synthesis of silver nanoparticles by laser ablation in pure water **806** 803–6

[11] Sadrolhosseini A R, Shukri A and Muhammad B 2013 Laser ablation synthesis and optical properties of copper nanoparticles 14–21

[12] Sadrolhosseini A R, Rashid S A, Zakaria A and Shameli K 2016 Green Fabrication of Copper Nanoparticles Dispersed in Walnut Oil Using Laser Ablation Technique **2016**

[13] Fernandes G D, Gómez-coca R B, Pérez-camino M C, Moreda W and Barrera-arellano D 2017 Chemical Characterization of Major and Minor Compounds of Nut Oils : Almond , Hazelnut , and Pecan Nut **2017** 5–7

[14] Easawi K, Nabil M, Abdallah T, Negm S and Talaat H 2012 Plasmonic Absorption Enhancement in Au/CdS Nanocomposite *World Acad. Sci. Eng. Technol.* **6** 63–6

[15] Abdallah T, El-Brolosy T A, Mohamed M B, Easawi K, Negm S and Talaat H 2012 Effect of shape and interstice on surface enhanced Raman scattering (SERS) of molecules adsorbed on gold nanoparticles in the near-dipole and quadrupole regions *J. Raman Spectrosc.* **43** 1924–30

[16] Donnell G O 2004 Particle Size Determination using Mie Scattering 33–4

[17] Yashunin D A and Korytin A I 2017 Localized Plasmon resonance in metal nanoparticles using Mie theory Localized Plasmon resonance in metal nanoparticles using Mie theory **2**

[18] Mafune F, Kohno J, Takeda Y and Kondow T 2000 Formation and Size Control of Silver Nanoparticles by Laser Ablation in Aqueous Solution 9111–7

[19] Desarkar H S, Kumbhakar P and Mitra A K 2012 Effect of ablation time and laser fluence on the optical properties of copper nano colloids prepared by laser ablation technique 285–91

[20] Zamiri R, Zakaria A, Ahangar H A and Sadrolhosseini A R 2010 Fabrication of Silver Nanoparticles Dispersed in Palm Oil Using Laser Ablation 4764–70

[21] Stafe M, Marcu A and Puscas N N *Pulsed Laser Ablation of Solids*.

Thermal Diffusivity and Thermal Effusivity Enhancement of Graphene based Polyvinyl Chloride Nanocomposites

T. A. El-Brolossy [1*], A. Almutawa [2]

[1] Physics Department, Faculty of Science, Ain shams University, Cairo, Egypt.
[2] Physics Department, Faculty of Science, King Faisal University, Al-Ahsa, Saudi Arabia.

* E-mail: elbrolosyta@gmail.com

Abstract. The development of thermal diffusivity, thermal effusivity and heat capacity per unit volume of polymer nanocomposites, based on PVC filled with different concentration of graphene flakes up to 1.3 vol.% have been investigated using photoacoustic technique. The composites were prepared in the form of thin films with relatively small thickness (23–30 μm). Morphological and structural characterizations of the prepared composite films have been performed using laser confocal microscope, SEM and Raman spectroscopy. The composites exhibit an improvement in both thermal diffusivity (175%) and thermal effusivity (136%) of their values for neat polymer at the highest graphene content used. However, with increasing graphene content no significant variations in the measured values of heat capacity were obtained. The experimental results were analyzed using the mixed model for the thermal conductivity of a two-phase system. We obtained good agreement between our results and the predictions of this model at thermal diffusivity (= $(1.5 \pm 0.07) \times 10^{-4}$ m^2s^{-1}) and thermal effusivity (= $(1.83 \pm 0.06) \times 10^4$ $Ws^{1/2}m^{-2}K^{-1}$) of fillers. The predicted thermal diffusivity value of graphene flakes is comparable to the published thermal diffusivity of graphene plates. On the other hand, according to the author's knowledge, the thermal effusivity of graphene was not previously studied. However, the thermal effusivity predicted in this work is in fair agreement with that estimated using the reported thermal properties data of graphene.

Keywords: Graphene flakes, polymer nanocomposites, photoacoustic, Thermal diffusivity, Thermal effusivity.

1. Introduction

Effective thermal managing in modern electronics devices has become of fundamental necessity. As the electronic components of these devices have become more integrated and miniaturized, the heat they generate could reduce the lifetime and efficiency of the device if not dissipated rapidly. Therefore, materials with considerable thermal conduction are necessary to remove the heat away and maintain the system [1]. Polymer nanocomposite with enhanced thermophysical properties can effectively dissipate such heat. Generally, polymers have various applications [2, 3] however most polymers have many defects due to their amorphous state that cause several phonon scatterings leading to a low thermal conductivity (0.1 - 0.5 $Wm^{-1}K^{-1}$) [4]. The progress of polymeric nanocomposites based on standard polymers and conductive carbon rich materials ,especially carbon nanotubes (CNTs), [5–9] is very promising way to obtain new materials with novel structural and functional properties surpass those of the neat components and other nanocomposite systems with other fillers [10–13]

Compared to CNTs, graphene sheets (GS) have higher surface-to-volume ratios due to the unreachability of the inner surface of the nanotubes to the polymer molecules. Also, the reported value of thermal conductivity of single layer graphene attains 4000 – 7000 W/m·K at room temperature which is higher than that of CNTs [14–16]. Furthermore, graphene is obtained from naturally occurring graphite suggesting that lighter and functional improved graphene-reinforced nanocomposites can be produced at lower cost. This renders graphene sheets possibly more promising for varying all matrix properties including thermal properties and thermal stability. In recent years, efforts are devoted towards graphene-based nanocomposites [17-19]. More recently, the thermal conductivity of graphene-based composites has made rapid progress [20–24]. However, the size and thickness of GSs show a discrepancy depending on how they are produced; accordingly, their reported thermal conductivity values vary considerably [25]. Also, due to the extremely large specific surface area of graphene, the number of interfaces as well as the interfacial bonding between graphene and the host polymer controls the interfacial thermal resistance and consequently the heat transfer process through the produced nanocomposite. Therefore, more research on graphene-based polymer nanocomposites of enhanced thermal properties is needed.

Not only thermal conductivity is the relevant thermal parameter concerning heat conduction in solids, but also both of thermal diffusivity (α) and thermal effusivity (e) are two highly significant thermal parameters that govern heat transfer and heat exchange processes for solid or liquid substances respectively. These quantities are given by $\alpha = k/\rho c$ and $e = \sqrt{k\rho c}$, where c is the specific heat capacity, ρ is the mass density while k is the thermal conductivity. Thermal diffusivity is a characteristic property measures the rate of heat diffusion within the material; it controls how fast heat will flow within the material. On the other hand, thermal effusivity is the ability of the material to exchange heat or it controls the heat transfer process at the material's interfaces. Therefore, it is related to thermal impedance at material's surface. Such thermophysical properties of graphene-based polymer nanocomposites are rarely studied or even not exist in literatures [26] .

In last couple of decades, optical measurement methods have been recognized as non-invasive and accurate techniques for thermal measurements [27- 29]. In this work we employ the photoacoustic (PA) technique as an optical technique to follow the development of both thermal diffusivity and thermal effusivity of Polyvinyl chloride/graphene nanocomposite with increasing graphene content.

2. Theoretical Aspects

The nonradiative relaxation of photoexcited states usually results in heating of the sample. The released heat energy carries information regarding both the absorbed energy and the thermal properties of the sample. A group of such spectroscopic methods based on the detection of photo-induced heating of the sample are called the photothermal methods. The PA technique is a photothermal detection method that has been attested [30, 31] to be a potent tool for studying optical and thermal properties of materials. The master process of PA technique is based on illuminating the sample placed in a closed chamber filled with a gas such as air with monochromatic radiation of any desired wavelength and modulated intensity with suitable acoustic frequency (*f*). The nonradiative decay of the absorbed radiation produces a periodic heat diffused from the sample to the air adjacent to the sample surface. This temperature variation creates a pressure fluctuation in the air within the cell which is detected as an acoustic signal by a sensitive microphone attached to the chamber. At fixed incident wavelength the PA signal becomes function of modulation frequency only. One of the important parameters that control the amplitude of the PA signal is the thermal diffusion length μ ($= [\alpha/\pi f]^{\frac{1}{2}}$). For an appropriate sample thickness l_s, one can obtain a cross over from a thermally thin regime ($\mu_s > l_s$), to a thermally thick regime ($\mu_s < l_s$) by increasing the modulation frequency *f*. According to Charpentier *et al.* [32] as well as Bennett and Patty [33] the PA signal amplitude Q for the case of an optically opaque sample can be expressed as

$$Q = A\left(\frac{1+R^2e^{-4x}+2\,Re^{-2x}\cos 2x}{x^2}\right)[(1-R^2e^{-4x})^2 + 4R^2e^{-4x}\sin^2 2\,x]^{-1/2} \quad (1)$$

With $x = {}^{l_s}/_{\mu_s} = [l_s(\frac{\pi f}{\alpha})^{\frac{1}{2}}]$

A is constant non-related to the sample and $R = (1-b)/(1+b)$ is the thermal wave reflection coefficient with b given by $b = e_b/e_s = (\kappa_b\rho_b c_b/\kappa_s\rho_s c_s)^{1/2}$. Here e_b and e_s are thermal effusivities of the backing material and the sample, respectively. Equation (1) can be used to estimate thermal diffusivity and thermal effusivity of solid samples under certain experimental conditions.

3. Experimental Setup

3.1 Materials

Polyvinyl Chloride (PVC) average molecular weight 1020 was purchased from Sabic, KSA and solvent Tetrahydrofuran (THF) stabilized with 0.1% quinine from LOBACHEMIE. Graphene was obtained from

SKYSPRING Nanomaterials Graphene Nano powder product 0540DX. Graphene nanopowder, product 0540DX are very thin (1 – 5 nm), flat particles with large diameters (< 2 μm). Owing to the flat shape of these particles, they are particularly sensitive to van der Waals attractive forces and have an affinity to re-aggregate in the dry state. For this reason, we got our nanopowder in a granular form. These granules are friable collections of individual particles that prevent agglomerations and are easily broken with mechanical agitation.

3.2 Preparation of PVC/GNF nanocomposites Samples

The method which was extensively used in dispersing of nano materials within thermoplastic resins is done by dispersing the particles in an aromatic solvent, dissolving the polymer in the same solvent, mixing the two solutions with required ratios and then evaporating the solvent to obtain the thermoplastic with an excellent dispersion of nanopowder. We have used this technique to prepare nanocomposite sheets of PVC with graphene flakes (PVC/GNF) of various GNF wt. % using THF as a solvent. In a typical process, PVC was first dissolved in 40 mL of THF with continuous stirring (for about 3 h) up to complete dissolution. The appropriate amount of graphene was dispersed in THF by ultrasonic vibrations for 10 minutes. The two solutions were mixed with continuous stirring for 24 h to form stable graphene/polymer solution. This confirms the proper dispersion with no detectable precipitation. The resulting solution will be poured to Petri dishes with fixed area and dried at room temperature. After solvent evaporation, the film was peeled off where its thickness was determined using digital micrometer. The prepared films were labeled PVC-Pure, PVC/GNF-0.1%, PVC/GNF-0.3%, PVC/GNF-0.5%, PVC/GNF-0.8%, PVC/GNF-1.0% and PVC/GNF-2.0%. Fig. (1) shows the photographic pictures of some prepared films. The films seem almost homogenous, with a clear increase in the darkness as GNF loading increases.

Pure PVC	**0.5 % GNF**	**2.0 % GNF**

Figure 1. Photographic pictures of the composite samples.

3.3. *Photoacoustic Experiment*

Photoacoustic technique in front surface illumination for optically opaque samples has been used to estimate thermal properties of composite samples. Samples are films prepared as described previously with thickness (~23 – 33 μm) measured using digital micrometer. Good adhesion (using silicon grease) to the quartz glass plates as a backing material ensuring good thermal contact and avoiding the appearance of extra thermal resistance. The used quartz plates are of circular shape (5 mm diameter and 1.5 mm thickness) with well-known thermal properties (e_q = 1503 $Ws^{1/2}\ m^{-2}\ K^{-1}$, $\alpha_q = 0.87x10^{-6}\ m^2/s$). To get rid of the effect of varying absorption energy of the samples and verify the required condition of strong PA signal, we quoted the sample surface with very thin layer of black ink. This ultra-thin thermal layer ensures the same absorption energy for all samples without any significant effect on the measured thermal properties of the sample. As a result, variations in the measured PA signal from one sample to the other depend only upon the variation in their thermal properties. The sample was then placed within the PA cell and illuminated with 514.5 nm line of Ar^+ laser as an exciting beam with about 20 mW powers. We used a mechanical chopper to chop the incident beam at the desired frequencies covering the range from thermally thin to thermally thick region for all samples. The PA signal amplitude was recorded using a dual channel digital lock-in amplifier (SR830). For further confirmation of the results the measurements were performed for more than one sample of the same type.

4. Characterization

4.1 LCM and SEM

The surface morphology of the prepared composites was investigated using laser confocal microscope (LCM) of the micro Raman model: Labram HR Evolution-Horiba as well as SEM. A confocal microscope creates sharp images of a specimen that would else appear hazy when viewed with a conventional microscope. The image is clear and of better contrast than that of a conventional microscope. This is illustrated in Fig. 2 (a-c) that shows the two-dimensional LCM images of 0.1% wt., 0.5% wt. and 2%wt. PVC/GNF composite samples respectively. The images provide clear difference in the surface morphology and particle distribution with increasing GNF loading. A reduction in the space between particles with increasing GNF content is evidently observed. Also, as the concentration increases the graphene plates tend to aggregate as clearly observed for the sample of the highest concentration.

The SEM image of the graphene powder using [Fe-SEM JSM -7600 Jeol. Japan] presented in Fig. 3 (a). The layered structure of the friable granular graphene is clearly observed. The image shows several thin layers stacked over each other. Also, the surface morphology of the prepared samples has been studied using SEM. Fig. 3 (b & c) shows SEM images of neat PVC and PVC/GNF 2% *wt.* respectively. The smooth and flat surface of the pure PVC is clearly observed. In contrast, the random distribution of hills in the composite image reveals that most of the graphene nano flakes are well dispersed in the PVC matrix, with a few restacks. It should be noted from the low contrast of the image that graphene flakes seem to be masked inside the polymer matrix and are well dispersed in the PVC.

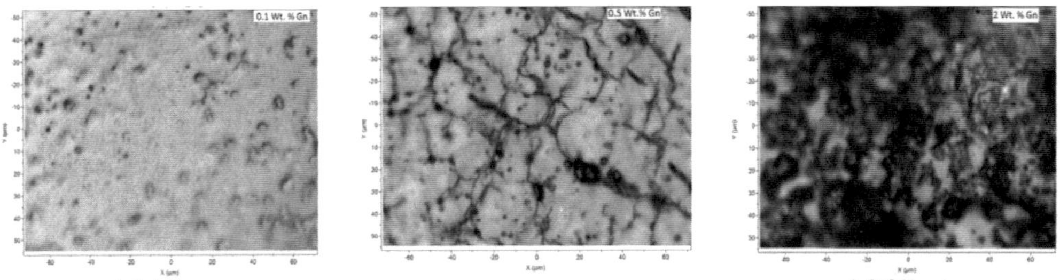

a) 0.1% wt b) 0.5% wt c) 2.0% wt

Figure 2. LCM images of PVC/GNF of different GNF loading as indicated

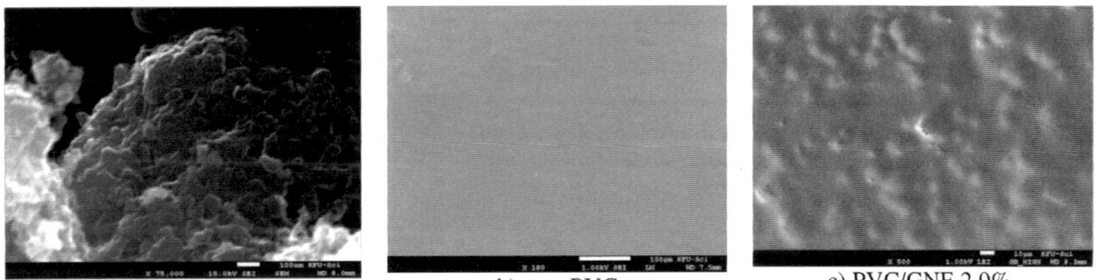

a) GNF b) neat PVC c) PVC/GNF-2.0%.

Figure 3. SEM images of graphene powder (a) neat PVC (b) and PVC/GNF 2% *wt.* (c)

4.2 Raman measurements

The Raman spectra of GNF sample and PVC/GNF nanocomposites have been obtained using the micro Raman spectrometer (model: LabRam HR Evolution-Horiba). The excitation source is He-Ne Laser of spot-size: 1-2 microns at 10% numerical density filter for 633 nm. Fig. 4 (a) shows Raman bands of Graphene powder in the range (800 – 2900 cm^{-1}). The spectrum shows the two characteristic bands, G band at (≈1575 cm^{-1}) and 2D band at (≈ 2645 cm^{-1}). The intensities and shapes of these bands reveal the multilayer structure of the graphene sample. Also, the appearance of a band at (≈ 1325 cm^{-1}) which is assigned as D band indicates the presence of defects and disorders in the graphene structure.

Raman scattering measurements of PVC/GNF composites were performed to investigate the structure as well as the distribution of GNF within the PVC matrix. Fig. 4 (b) shows the development of the G band of the GNF within the composite of different graphene *wt. %*. It is readily observed that the intensity of G band increases with increasing CNF contents. This may be considered as an argument of fair distribution of GNF within the polymer matrix. Also, the obtained G band of GNF within the polymer is (≈1587± 3cm^{-1}) for 2 % *wt*,

(≈1588± 3cm^{-1}) 0.8 % *wt.* and (≈1592 ± 3cm^{-1}) for 0.5 % *wt* which are blue shifted relative to that of the granular graphene. This blue shift indicates a decrease in the number of stacked graphene layers within the polymer matrix with a tendency to some agglomeration with increasing loading percentage.

Fig. 4(c) shows 2D band at (≈ 2664 cm^{-1}) of GNF in 2% *wt.* sample. We couldn't detect this band in the samples of lower graphene content. Also, this band is blue shifted as well as narrower (smaller FWHM) from that of the graphene powder. These observations confirm the reduction of the number of stacked layers of GNF within the polymer matrix relative to that of graphene powder. This agrees with the nature of our graphene sample which is supplied in a friable granular form (collection of multiple particles). With ultrasonic vibrations during composites' preparation, these particles were separated and distributed within the polymer matrix.

As it was mentioned before, the D band is associated to the defects and disorders in the graphene structure. Fig. 4(d) shows Raman spectrum of 2% *wt.* sample in the range 1200 cm^{-1} – 1800 cm^{-1}. The two graphene bands G and D together with the C – H ant symmetric stretching of PVC was detected. When we compare this spectrum with that of graphene powder (Fig. 3(a) it is directly noticed that the intensity ratio I_D/I_G of PVC/GNF composite of 2% *wt.* is larger than that of graphene powder. Obviously, this means an increase in the defects and/or disorder in the GNF within the PVC matrix relative to the graphene powder. This result suggests interfacial interaction between GNF and PVC.

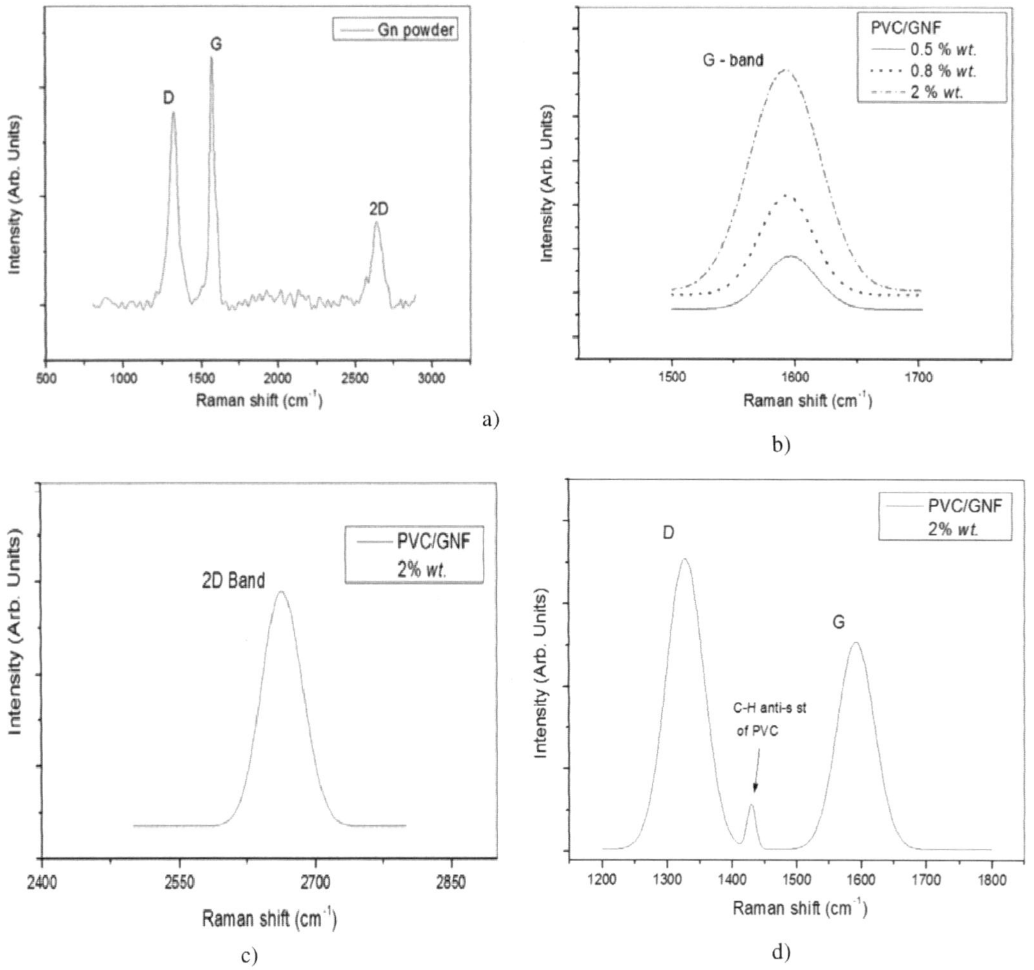

Figure 4. Raman spectra of GNF powder (a), G band of GNF of different concentrations within the PVC matrix as indicated (b), 2D band of 2% *wt* PVC/GNF sample (c) and D and G bands of 2% *wt* PVC/GNF sample (d)

IOP Conf. Series: Materials Science and Engineering **762** (2020) 012006 doi:10.1088/1757-899X/762/1/012006

5. Results and Discussion

5.1 Thermal diffusivity and thermal effusivity measurement

A distinctive variation of the PA amplitude with chopping frequency in the range (4 – 700 Hz) in the double logarithmic presentation of the PVC film of thickness 29 μm is presented in Fig. 5(a). Solid line represents the best fit of the experimental data (scattered points) according to the Eq. (1). One can clearly see in Fig. 5 (a) the varying slope at higher frequencies indicating a transfer from thermally thin to thermally thick sample. This contrasted to the straight-line behavior (constant slope) for the thermally thick sample (of thickness 350 μm) shown in the inset of Fig. 5 (a). By the aid of the known thermal effusivity of the backing material and the measured thickness of the sample we obtained fitting curves by fitting the three parameters α, e, and A to Eq. (1). The estimated values are α = 1.28 x 10^{-7} m^2/s and e = 450 $Ws^{1/2}$ m^{-2} K^{-1} accordingly, the thermal conductivity $k(e\sqrt{\alpha})$ is 0.161 Wm^{-1} K^{-1}. These values agree well with the reported values of PVC thermal properties [34] which indicates the suitability of the method for such measurements .

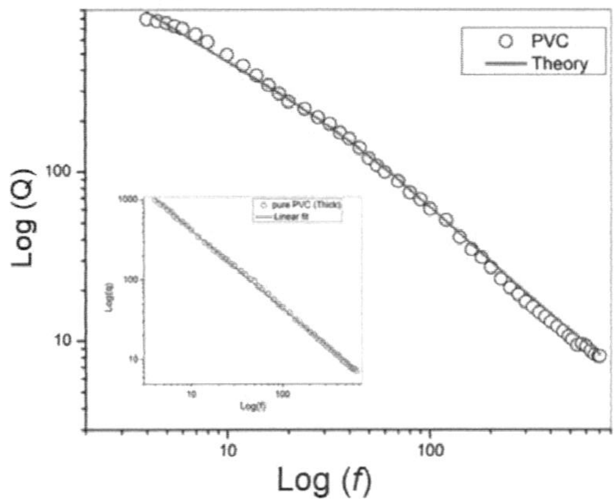

Figure 5. log – log plot of Q *vs.* f for pure PVC (thin sample 0.029 mm). The inset is the same relation for a thick sample

PA signal amplitude of composite samples of known thicknesses and different GNF weight content (0.1, 0.3, 0.5, 0.8, 1, and 2 wt. %) were measured. Typical variations of the PA signal amplitude with chopping frequency in the double logarithmic presentation for composite samples of (0.1 and 1 wt. %) are shown in Fig. 6 (a & b) respectively. Solid line represents linear fitting of the experimental data (scattered points) according to Eq. (1) as performed before. It is obvious that the behavior of amplitude variations with the chopping frequency is in good agreement with the theory for the two samples. The PA signal of other composites (not shown) revealed similar behavior

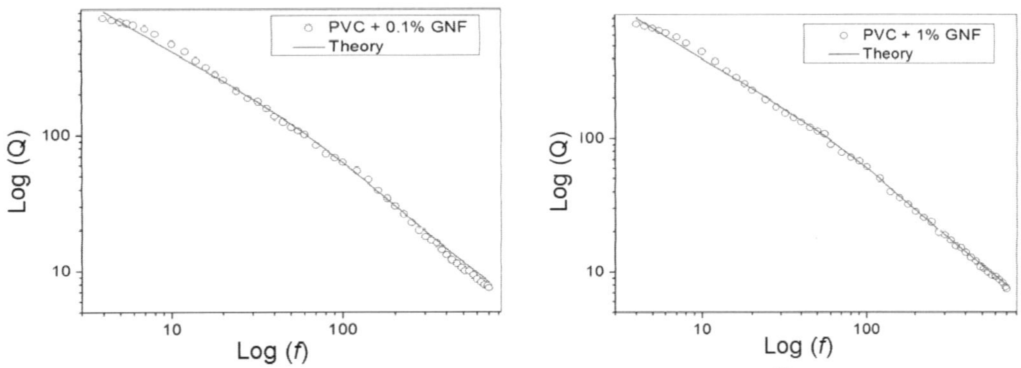

a) b)

Figure 6. log – log plot of Q *vs.* f for PVC/GNF 0.1 wt. % (a) and 1 wt. % (b)

The estimated values of α of the PVC/GNF composites with different GNF weight percentage, show significant enhancement with increasing GNF content reaches to about 175% of its original value α_o of pure PVC at the highest GNF content used (2 *wt.* %). This signifies that addition of GNF to PVC provides a new way for heat diffusion and the composite becomes better thermal diffuser than the neat polymer. Also, the obtained values of the thermal effusivity reveal an enhancement with increasing the GNF content. However, in comparison with the results obtained for α, improvement of e is lower reaches to about 131.7% of its original value e_o at the highest GNF content as shown in Fig. 7.

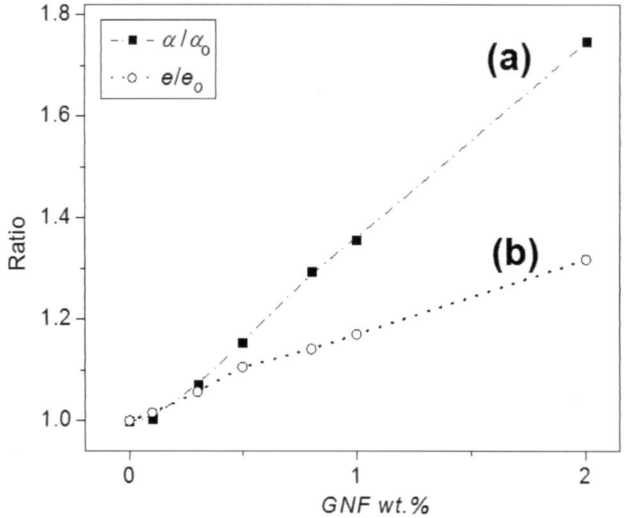

Figure 7. Ratio of thermal diffusivity of PVC/GNF to that of pure PVC and (b) Ratio of thermal effusivity of PVC/GNF relative to that of pure PVC vs. *wt.* % of GNF

Now, we are going to investigate the development of these thermal properties versus the volume percentage of each individual phase of the composite in order to estimate the thermal properties of the filler itself. The GNF volume fraction v in the composites was calculated and found to be within the range 0 – 1.3 *vol.* %. For such type of mixed material, the thermal properties are discontinuous functions of the position, therefore neither Fourier's law, nor the heat conduction equation are applicable. Consequently, the concept of 'effective' properties, that means the properties of the equivalent homogeneous material which produces the same physical effects, appears to be suitable. The effective heat capacity at most times follows the mixture rule. Thus, for a two phase composite the effective heat capacity per unit volume is given by [35]

$$(\rho c)_{eff} = v_1(\rho c)_1 + v_2(\rho c)_2 \quad (2)$$

Where v_i is the *vol.* % of component *i*. As can be seen, it only depends on the volume fraction of the components. Using the measured values of both (α) and (e), $(\rho c)_{eff}$ can be determined using the relation

$$(\rho c)_{eff} = {}^{e}/_{\sqrt{\alpha}} \quad (3)$$

Fig. 8 shows the measured values of the effective heat capacity per unit volume of PVC/GNF composites (scattered points) together with the calculated values (solid line) using Eq. (2) versus *vol.* % of GNF. Clearly, the calculated values of $(\rho c)_{eff}$ are nearly constant with increasing v in our range which is predicted as both ρ and c varies with v in an opposite manner. Wherein ρ of GNF (2.1 g/cm^{-3}) is larger than that of PVC (1.39 g/cm^{-3}) [34] while c of GNF (≈0.7 Jg^{-1}K^{-1}) [36] is lower than that of PVC. The experimental results reasonably agree with these calculated values, where no significant variations in the measured values of $(\rho c)_{eff}$(within the experimental uncertainties) were obtained as can be noticed. This may be a sign of the adequate dispersion of GNF in our PVC matrices. Some authors [37] obtained an unpredicted behavior with addition of MWCNTs which caused a specific heat capacity enhancement of the polymer composites. They ascribed it to the ineffective dispersion of nanotubes within the polymer matrix.

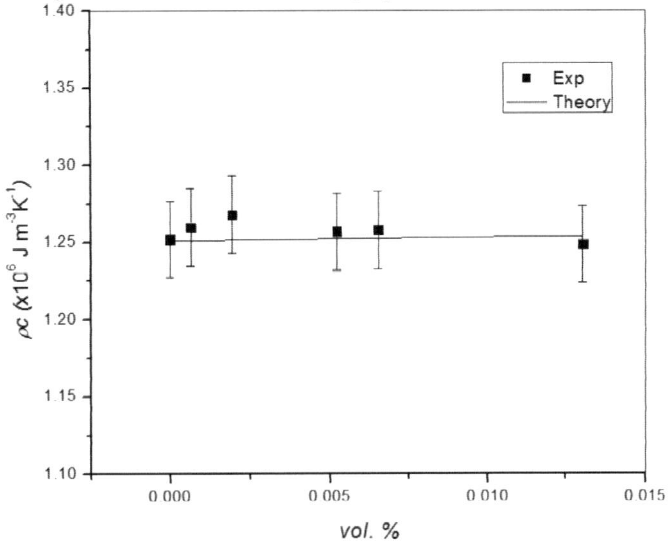

Figure 8. Volume heat capacity (ρc) as a function of GNF *vol. %*

Also, the invariance of ρc with GNF addition suggests that the increase of both thermal diffusivity and thermal effusivity of PVC/GNF composites with increasing GNF loading is essentially due to the enhancement of thermal conductivity. This might allow some of the theoretical and/or empirical models that forecast thermal conductivities of two-phase systems to be valid for both thermal quantities (diffusivity and effusivity).

5.2 Thermal Properties of GNF

We make full use of the mixed empirical model of polymer composites that was based on the parallel and series models of thermal conductivity of two phase systems [38] to foretell both thermal diffusivity and thermal effusivity of the composite samples. This model is given as [9]:

$$\lambda_E^{\zeta} = v\lambda_F^{\zeta} + (1 - v)\lambda_P^{\zeta} \quad (4)$$

Where λ represents the thermal quantity (either α or *e*); the sub-indices *E*, *F* and *P* indicate to composite, filler (GNF) and neat polymer respectively and ζ is a constant with values between −1 and +1 based on the experiment.

The predictions of this model fit the measured thermal diffusivity of the composites as shown in Fig.9(a) at acceptable value of ζ= 0.44 and thermal diffusivity of GNF (α_{GN} = (1.5 ± 0.07) × 10^{-4} m^2s^{-1}). The obtained thermal diffusivity value of GNF is comparable to the values measured by Cabrera, H., et al. [39] for thermal diffusivity of graphene plates. Conversely, the model fits thermal effusivity of the composites as shown in Fig. 9(b) at ζ= 0.82 and thermal effusivity of GNF (= (1.83 ± 0.06) × 10^4 $Ws^{1/2}m^{-2}K^{-1}$). The comparison with the reported values of GNF thermal effusivity could not be made as such values are not exist in the literature. However, the calculated thermal effusivity using the estimated value of α_{GN} as well as the reported values of both ρ and *c* for GNF through the relation $e = \rho c\sqrt{\alpha}$ yields a value (1.8 × 10^4 $Ws^{1/2}m^{-2}K^{-1}$) which is close to the predicted one.

The above results reflect the success of this indirect method to estimate thermal diffusivity and thermal effusivity of GNF. However, the predicted values of these thermal properties especially α lie in the lower limit of the reported values, this might be due to number of graphene layers in our GNF as well as the thermal barrier layer usually exists at graphene/polymer interface, which effectively reduces the heat diffusion in composites.

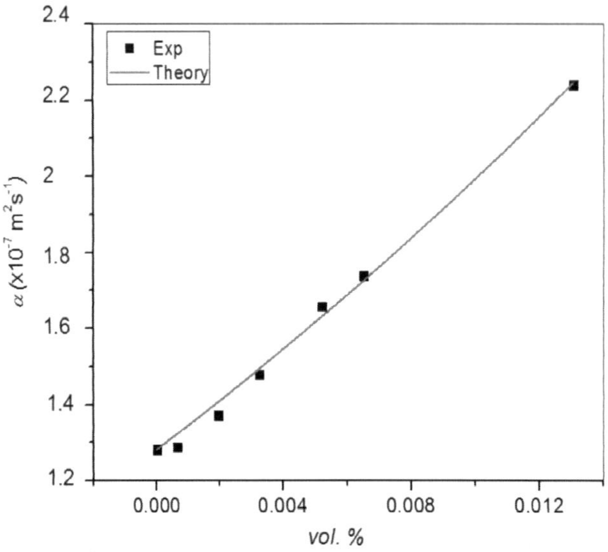

a)

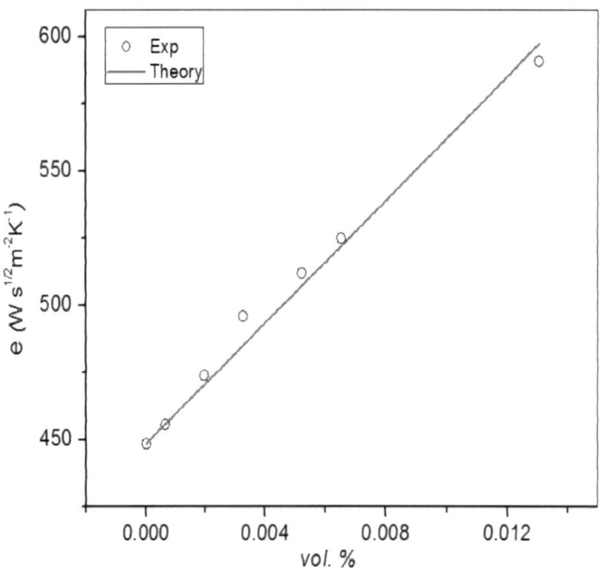

b)

Figure (9): Thermal diffusivity (a) and effusivity (b) of PVC/GNF as a function of GNF *vol. %*. Solid lines represent calculated results using equation (4)

6. Conclusion

Photoacoustic experiment in the front surface illumination has been employed to measure thermophysical properties, including thermal diffusivity, thermal effusivity and volumetric heat capacity of PVC/GNF nanocomposites of varying GNF content (0 – 1.3 *vol. %*). Continuous enhancement of both (α) and (e) of PVC was obtained via the inclusion of GNF, which attained 1.75 times the thermal diffusivity and about 1.36 times the thermal effusivity at 1.3 vol.%. The normal heat capacity behavior of PVC/GNF composites together with the gradual increase of the G band intensity with increasing GNF content suggest enough dispersion of GNF in our PVC matrices. Using the mixed model for thermal conductivity of a two-phase system, we were able to estimate acceptable values of thermal diffusivity and thermal effusivity of GNF. These results confirm the appropriateness of this optical method to measure thermal properties of polymer nanocomposites.

References

[1] Chen, H., Ginzburg, V.V., Yang, J., Yang, Y., Liu, W., Huang, Y., Du, L., Chen, B. (2016) Prog. Polym. Sci., 59, 41–85.
[2] Chattopadhyay, S., Madras, G. (2002) Polymer degradation and Stability. 78, 519–524.
[3] Adel, R., Abdallah, T., Moustafa, Y.M., Al-Sbagh, A.M., Talaat, H. (2015) Superlattices and Microstructure. 86, 62–67
[4] Huang, X., Jiang, P., Tanaka, T. (2011) IEEE Electr. Insul. Mag., 27, 8–16.
[5] Song, Y.S., Youn, J.R. (2006) Carbon, 44, 710–717.
[6] Moisala, A., Li, Q., Kinloch, I.A., Windle, A.H. (2006) Compos. Sci. Technol., 66, 1285–1288.
[7] Aljaafari, A., Ibrahim, S., El-Brolossy, T. (2011) Composites Part A, 42(4), 394-399.
[8] El-Brolossy, T.A. Ibrahim, S.S., Alkhudhayr, E.A. (2015) Polymer composites, 36(7), 1242-1248.
[9] El-Brolossy T.A., Ibrahim, S. S. (2014) Phys. Scr, 89, 105701 (9pp).
[10] Choi, S., Kim, J. (2013) Compos. B Eng., 51, 140–147.
[11] Li, B., Li, R., Xie, Y. (2017) J. Mater. Sci., 52, 2524–2533.
[12] Im, H., Kim, J. (2011) J. Mater. Sci., 46, 6571–6580.
[13] El-Brolossy, T.A., Ibrahim *S.S.* (2010) Thermochimica Acta 509, 46–49.
[14] Balandin, A.A. (2011) *Nat Mater.* 10, 569–81.
[15] Balandin, A.A., Ghosh, S., Bao, W, Calizo, I., Teweldebrhan, D., Miao, F., Lau, C.N. (2008) *Nano Letters* 8 (3), 902-907.
[16] Ghosh, S., Calizo, I., Teweldebrhan, D., Pokatilov, E.P., Nika, D.L., Balandin, A.A., Bao, W., Miao, F., Lau, C.N. (2008) Appl Phys Lett. 92**,** 151911.
[17] Mukhopadhyay, P., Gupta, R.K. (2012) *Graphite, Graphene, and their polymer nanocomposites*. CRC Press.
[18] Kuila, T., Bose, S., Mishra, A.K. Khanra, P., Kim, N.H., Lee, J.H. (2012) Progress in Materials Science,. **57**(7), 1061-1105.
[19] Regueira, R., Suckeveriene, R.Y., Brookl, I., Mechrez, G., Tchoudakov, R., Narkis, M. Graphene. (2015) **4**(01), 7-19.
[20] Burger, N., Laachachi, A., Ferriol, M., Lutz, M., Toniazzo, V., Ruch, D. (2016) Progress in Polymer Science. 61, 1–28.
[21] Shtein, M., Nadiv, R., Buzaglo, M., Kahil, K., Regev, O. (2015) Chem. Mater., 27, 2100–2106.
[22] Li, A., Zhang, C., Zhang, Y-F. (2017) Polymers, 9, 437.
[23] Song, W-L., Wang, W., Veca, L.M., Kong, C.Y., Cao, M-S., Wang, P., Meziani, M.J., Qian, H., LeCroy, G.E., Caoa, L. Sun, Y-P. (2012) J. Mater. Chem., 22, 17133.
[24] Zhang, C., Hao, X. L., Wang, C. X., Wei, N., & Rabczuk, T. (2017). *Scientific reports*, *7*, 41398.
[25] Noh, Y., Joh, H., Yu, J., Hwang, S.L., Lee, S., Lee, C.H., Kim, S.Y.,Youn, J.R. (2015) Sci. Rep. **5,** 9141.
[26] Gresil, M., Wang, Z., Poutrel, Q., Soutis, C. (2017) Sci Rep **7,** 5536.
[27] Ali, F.M., Yunus, W.M.M., Moksin, M.M., Talib, Z.A. (2010) Rev. Sci. Instrum. 81, 074901.
[28] El-Brolossy, T.A., Saber, O. (2013) Experimental Thermal and Fluid Science 44, 498–503.
[29] El-Brolossy, T.A., Saber, O., Ibrahim, S.S. (2013) Chin. Phys. B. 22(7), 074401.
[30] El-Brolossy, T.A., Abdallah, S., Abdallah, T., Mohamed M.B., Negm, S., Talaat, H. (2008) Eur. Phys. J. Special Topics. 153, 365–368.
[31] El-Brolossy, T.A., Abdallah, S., Hassanein, O.E., Negm, S., Talaat, H. (2005) Journal De Physicque IV 125.
[32] Helander, P. (1982) J. Photoacoustics. 1, 103-120.
[33] Chirtoc, M., Tosa, V., Bicanic, D., Torfs, P. Berichte der Bunsengesellschaft für physikalische Chemie, (1991). 95(7). 766-769.
[34] Mamunya, Y., Boudenne, A., Lebovka, N., Ibos, L. , Candau, Y. Lisunova, M. (2008) Composites Science and Technology,. 68(9), 1981-1988.

[35] Salazar, A. (2003) European journal of physics. 24(4), 351.

[36] Pop, E., Varshney, V., Roy, A.K. (2012) MRS bulletin. **37**(12), 1273-1281.

[37] Lizundia, E., Oleaga, A., Salazar, A., Sarasua, J-R. (2012) Polymer. 53(12), 2412-2421.

[38] Shen, M., Cui, Y., He, J., Zhang, Y-M. (2011) International Journal of Minerals, Metallurgy, and Materials. 18(5), 623-631.

[39] Cabrera, H., Mendoza, D., Benítez, J.L., Flores, C.B., Alvarado, S., Marín, E. (2015) Journal of Physics D: Applied Physics. 48(46), 465501.

Effect of CdS quantum dots size on Thermal and photovoltaic parameters of quantum dots sensitized solar cells

A. Khalid[1] K. Easawi[1], S. Abdallah[1] , M.G.EL-Shaarawy[2] , S. Negm[1], H. Talaat[3*]

[1] Department of Mathematical and Physical Engineering , Faculty of Engineering (Shoubra), Banha University, Cairo, Egypt

[2] Physics Department, Faculty of Science, Benha University

[3] Physics Department, Faculty of Science, Ain Shams University, Abbassia, Cairo, Egypt

* E-mail: : hassantalaat@hotmail.com

Abstract. The effect size of CdS Quantum Dots (QDs) on thermal and photovoltaic parameters is investigated. CdS QDs were adsorbed onto TiO_2 electrodes using successive ionic layer adsorption and reaction (SILAR) to act as sensitizers of quantum dots solar cells (QDSSCs). The CdS QDs sizes are estimated using optical absorption spectra and application of effective mass approximation model (EMA) as well as high resolution transmission microscopes (HRTEM). The ratio of TiO_2/CdS was confirmed by energy dispersive X-ray spectroscopy (EDX). Thermal parameters (thermal diffusivity α, thermal effusivity e and thermal conductivity k are measured by using Photoacoustic (PA) Technique. The photovoltaic parameters (open circuit voltage V_{oc}, short circuit current density J_{sc}, fill factor FF and efficiency η) of the assembled CdS QDs sensitized solar cells (QDSSCs) were determined under a solar illumination of 100 mW/cm2 (AM 1.5 conditions. Our results show that both the effective thermal conductivity k_{eff} and efficiency η of CdS QDs deposited on TiO_2 increase as the size of CdS QDs is increased.

Keywords: CdS QD; photoacoustic; SILAR; thermal properties

1. Introduction

The unique electronic and optical properties of semiconductor (QDs) have attracted significant attention to be used in QDSSC due to their large dipole moments and strong light absorption [1,2]. The dependence of their band gaps on their size makes them attractive candidates as sensitizers in photovoltaic solar cells because of the ability of tuning their band gap to the sunlight spectra [3, 4]. A regular QDSSC contains working electrode (TiO_2/QDs), redox electrolyte, and counter electrode [5, 6].Successive ionic layer adsorption and reaction (SILAR) technique are among the better known methods used to synthesize QDs [7- 9]. SILAR is used in our study since it provides a good quality surface spread with the capability of tuning energy band gap through controlling QDs size with low costs technique [10-13]. Furthermore, it can simply adjust the thickness by controlling the number of cycles. Many studies were carried out using semiconductor QDs (especially CdS) [14] as sensitizers to fabricate QDSSCs. The photovoltaic performance of QDSSCs based on SILAR as QDs deposition technique is remarkably enhanced compared with those using the direct method. .Measurement of thermal parameters such as thermal diffusivity (α), thermal effusivity (e) and thermal conductivity (k) for solar cell electrode are very essential to investigate the effect of rate of heat flow through the solar cell materials. More specifically K of the solar cell electrode is effect of operating temperature of solar cell. Photoacoustic technique (PA) is a photothermal detection technique; that is proved to be a powerful tool to study the optical and thermal properties of such Materials without particular sample treatment in a nondestructive manner [15-17]. In PA technique only the absorbed light energy contributes in the form of heat, while the scattered and reflected parts of light don't contribute

to the PT signals. Accordingly, PT spectroscopy more accurately measures optical absorption and thermal properties in highly scattering QDSSCs. Modulating of incident photon intensity causes temperature pulsations and the resulting temperature variation at the surface causes pressure variation in the PA cell [18]. Such pressure fluctuations can be detected by a sensitive microphone and constitute PA signal.
In this work CdS QDs were synthesized onto TiO_2 NPs by SILAR technique through increasing number of cycles under ambient conditions. The particles sizes were determined using (EMA) as well as HRTEM. The ratio of TiO_2/CdS was confirmed by (EDX). The size dependence of α, e and k for CdS QDSSs working electrode has been measured using PA technique. Also, the photovoltaic characteristic of the QDSSs (short current density J_{sc}, open circuit voltage V_{oc}, fills factor FF and efficiency for energy conversion η) are measured as a function of the varying size of CdS QDs.

2. Experimental

2.1. Deposition of the CdS QDs onto TiO_2 NPs photoelectrode

A colloidal paste of TiO_2 NPs was prepared according to our previous work [19]. The TiO_2 paste was deposited on a conducting glass substrate of Florien Tin Oxide SnO2:F (FTO) with sheet resistance of 7 Ω/sq and > 80% transmittance in the visible region, using a simple doctor blade technique. This was followed by annealing at 450 ◦C for 30 min and the final thickness was 8 μm after the solvent evaporation.
The TiO_2 photoanodes were sensitized with CdS QDs using SILAR method [20]. Where the photoanode is immersed for one min in 0.1 M ($Cd(NO_3)_2.H_2O$) in methanol as a cation source (Cd^{+2}) to be adsorbed on the TiO_2 electrode. The TiO_2 electrode was then rinsed with methanol to remove the excess Cd^{+2} cations. Then the photoanode is immersed in 0.1 M Na_2S methanolic solution as the anion source (S^{-2}) for one min in order to react with Cd^{+2} cations followed by methanol rinsing. This procedure constitutes as one SILAR cycle. Five samples obtained by increasing number of layer thickness of cycles were labeled a, b, c, d and e for 2, 4, 6, 8 and 10 cycles, respectively. The counter electrodes were prepared by coating another FTO substrate of resistance of 7 Ω/sq with platinum (Pt).

2. 2. Assembly of QDSSCs

QDSSCs were built as shown in Figure 1 by assembling the TiO_2 photoanodes sensitized with CdS QDs, aqueous polysulfide electrolyte and the counter electrodes. Parafilm was used as spacer between the two electrodes and also as a sealing to prevent the evaporation of the electrolyte. The space between the two electrodes was filled by polysulfide electrolyte which was prepared by dissolving 1.1M Na2S, 1.2 M S, and 0.12 M KCl in deionized water. The counter electrodes were prepared by coating another FTO substrate of resistance of 7 Ω/sq with platinum (Pt).

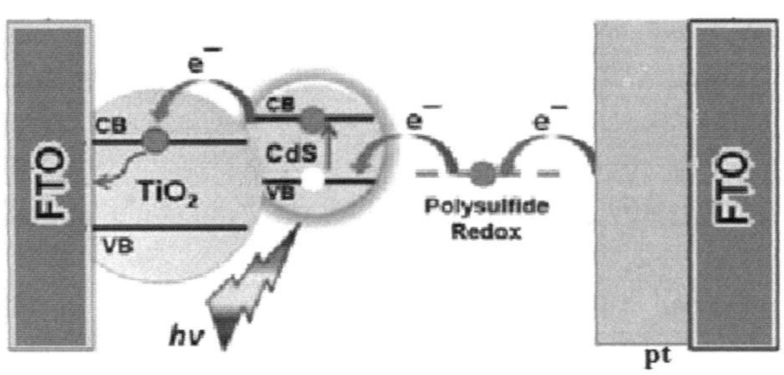

Figure1. Schematic representation of the assembled QDSSCs

2.3. Measurements

The optical absorption spectra of the working electrode (TiO_2/CdS) were measured using a UV–Visible spectrophotometer (JASCO V-670). The particle size of sample was measured using transmission electron microscope (TEM) (JEOL JEM-2100). The PA Thermal measurements were carried out using PA cell with a gas microphone detection technique and argon ion laser 6 W (514 nm) . The PA signal generated from the sample was amplified by a low noise preamplifier and further processed using a lock-in amplifier (Stanford Research System, Model SR830 DSP). A personal computer was interfaced to the system for automatic data acquisition and analysis

The current density voltage (J–V) characteristics are recorded with a Keithley SMU 2400 voltage source ammeter. The CdS QDSSCs were subjected to the irradiation of a solar simulator (San-Ei Electric XES-40S1) operating at 100 mW/cm^2 (AM 1.5G). The incident solar intensity illumination was adjusted to 1 sun condition using a Leybold certified silicon reference solar cell (Model: Bunkoukeiki BS-500BK). A J-V.

Characteristic curves of CdS QDSSCs with five particle sizes were obtained and all experiments were carried out under ambient conditions. The photoelectric conversion efficiency η was determined according to:

$$\eta = \frac{P_{max}}{P_{in}}\% = \frac{V_m I_m}{P_{in}}\% \qquad (1)$$

Where V_m and I_m are the maximum voltage and current respectively.

The fill factor FF is given by :

$$FF = \frac{V_m I_m}{V_{oc} I_{sc}}\% \qquad (2)$$

3. Results and Discussion

3.1. Optical measurements of CdS QD

TiO_2/CdS layer in ethanol was scratched from FTO coated substrate without disturbing FTO layer and placed on Cu-coated carbon grid for TEM measurement. Figure 2 shows the TEM image of TiO_2/CdS for sample (e) , the average particle size for CdS QD were estimated to be approximately 5.4 nm and average particle size for TiO_2 is 22 nm.

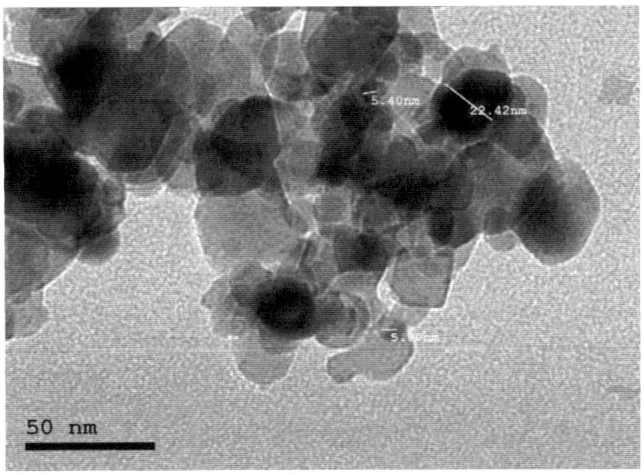

Figure 2. TEM micrographs of TiO_2/ CdS for sample (e)

Figure 3 show UV-visible absorption spectra for FTO/TiO_2 and FTO/TiO_2 / CdS photoanode with increasing number of SILAR cycles. It can be seen that the TiO_2 layer shows an absorbance edge at 400 nm corresponding to 3.1 eV band gap for TiO_2 nanoparticles. Also, there is a red shift of absorbance edge with increasing number of cycle. Samples a, b, c, d and e has peak absorption are 401 nm, 414 nm, 431 nm, 451nm and 465 nm for, respectively.

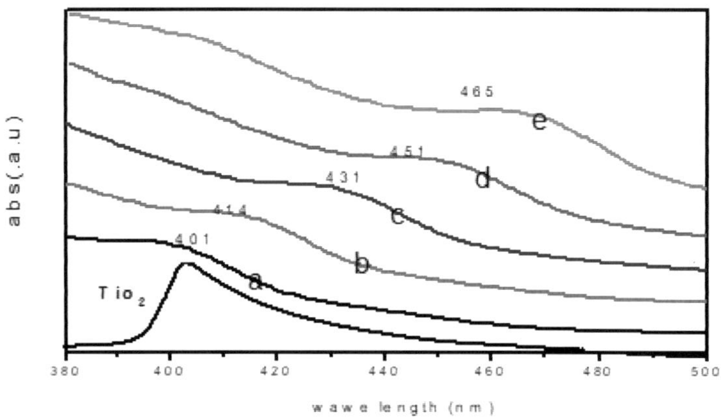

Figure 3. UV –Visible absorption spectra for TiO_2 and five TiO_2/CdS QDs samples with different number of cycles by SILAR method

The corresponding particle sizes of CdS QD are calculated using the effective mass approximation (EMA) [21].

$$E_{gn}(R) = E_{gb} + \frac{h^2}{8m^* R^2} - \frac{1.8e^2}{4\pi\varepsilon\varepsilon_0 R} \qquad (3)$$

where E_{gb} (2.4 ev) is the bulk CdS band gap , E_{gn} is the nanoparticle band gap , R is the radius of the CdS quantum dots , m^* (7.62×10^{-32} kg) is the reduced electron-hole mass, and ε (7.1) is the dielectric constant for CdS[22] . The obtained values of CdS particle sizes are 3.21 nm ,3.4 nm ,3.7 nm ,4.01nm and 4.4 nm for samples a ,b, c, d and e, respectively in close agreement with that obtained by TEM measurements. EDX determined by (JEOL JED-2300). Figure 4 shows the EDX spectra of TiO_2/CdS QDs sensitized working electrode where Cd, S, Ti and O atoms are indicated. The adsorption of CdS QDs onto the TiO_2 working electrode was confirmed by EDX.

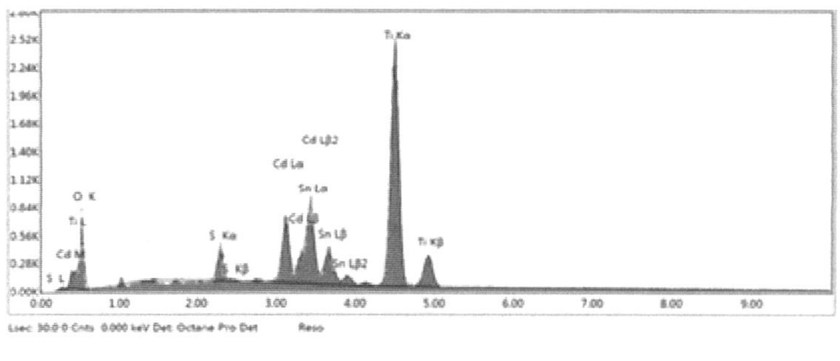

Figure 4. EDX of TiO_2/ CdS QD working electrode for sample e

3.2. Thermal measurments:-

The PA technique was employed to investigate the effective thermal parameter of (FTO/TiO_2/CdS QDs) photoanode that are important for solar energy conversion. Assuming that this photoanode as effectively FTO/TiO_2 and CdS QDs. The PA signal amplitude was recorded at various chopping frequencies (f) for each of the size photoanodes. The change of ln PA amplitude is plotted versus ln f for FTO/TiO_2 substrate is seen in Figure 5. A distinct change in the slope is clearly observed at a particular frequency f_c where the sample changes from being thermally thick to thermally thin where the crossover takes place [23].

The thermal diffusivity (α) of working electrode is then calculated using the relation [24].

$$\alpha_{eff} = f_c L^2 \quad \text{in} \quad m^2 / s \qquad (4)$$

where L is the working electrode thickness. The obtained value D is about $3.2\text{x}10^{-5}$ m^2/s. The plots of ln PA amplitude versus the ln f for photoanode (FTO/TiO_2/CdS QDs) of different CdS QDs sizes are shown in Figure 6(a) ,6.b, 6.c, 6.d and 6.e for sample a ,b, c, d and e, respectively. The calculated values of α_{eff} are displayed in Table 1. The results showed that, there is an increase of α_{eff} with increasing CdS QDs sizes which was about 216 % from its original value for the highest CdS QDs size.

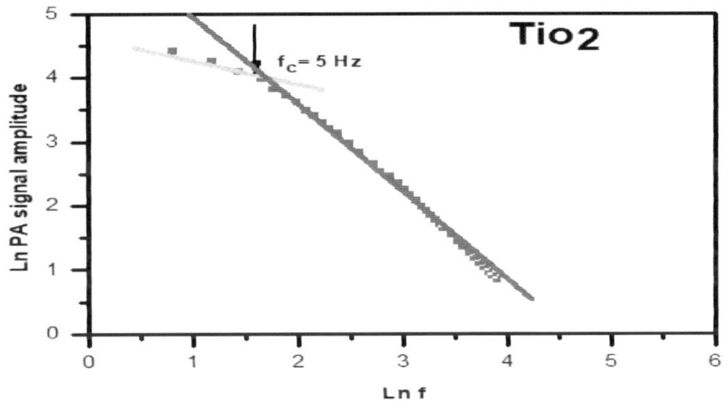

Figure 5. Variation of ln PA signal amplitude verses ln f for FTO/TiO_2.

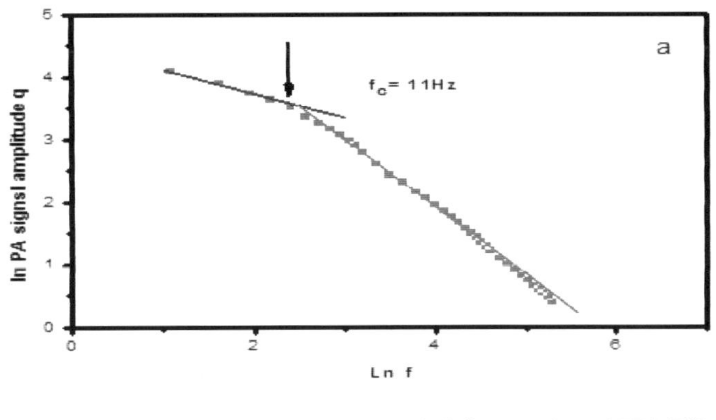

Figure 6.a Variation of ln PA signal amplitude verses ln f for sample a of FTO/TiO_2/CdS QDs.

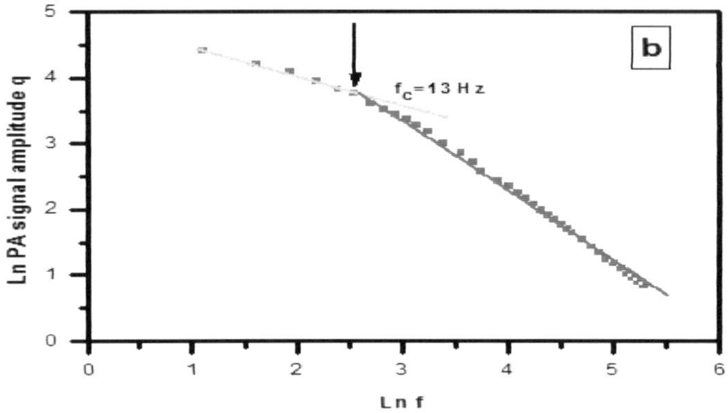

Figure 6.b Variation of ln PA signal amplitude verses ln f for sample b of FTO/TiO_2/CdS QDs.

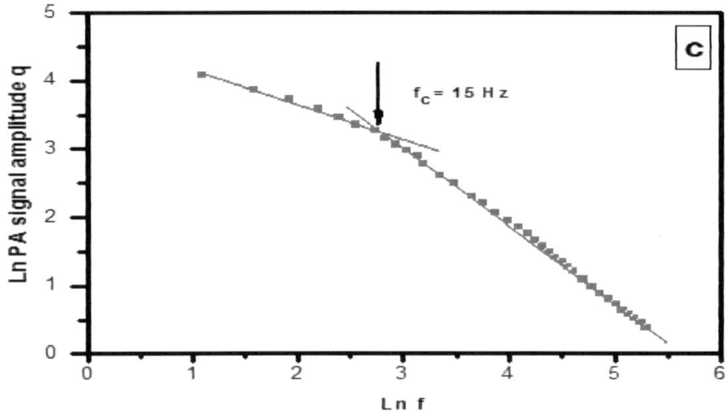

Figure 6.c Variation of ln PA signal amplitude verses ln f for sample c of FTO/TiO_2/CdS QDs.

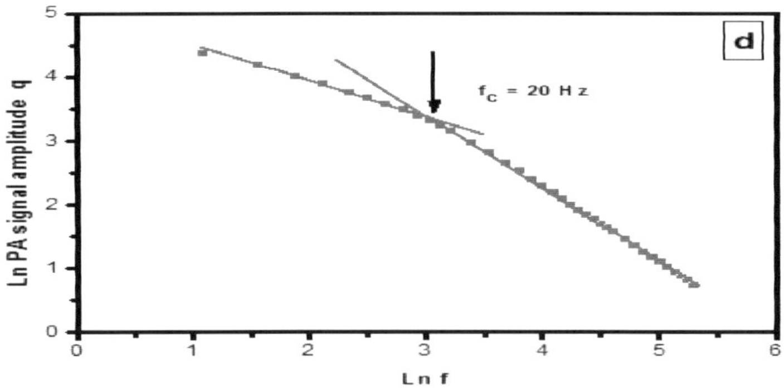

Figure 6.d Variation of ln PA signal amplitude verses ln f for sample e of FTO/TiO_2/CdS QDs.

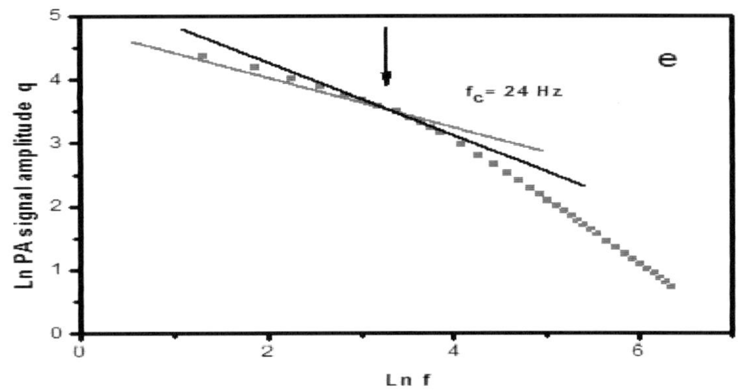

Figure 6.e Variation of ln PA signal amplitude verses ln f for sample e of FTO/TiO_2/CdS QDs.

The plots of ln (PA) against 1/ f for the same photoanode are seen in Figure 7 .The effective thermal effusivity e_{eff} were also estimated for our samples which are, thermally thick , hence the signal amplitude of PA is given by [25].

$$q = \frac{A}{e}\frac{1}{f} \quad (5)$$

where $$A = \frac{I_0 \gamma P_0 \alpha_g^{1/2}}{4\pi l_g T_0}$$

I_0 is the incident radiation intensity; T_0 and P_0 are the ambient pressure and temperature respectively, γ is gas specific heats ratio, α_g is the gas thermal diffusivity and l_g is the length of the gas column of the PA cell. Using the Si as a standard material of known effusiveity (=16061 $W.s^{1/2}.m^{-2}.K^{-1}$)[18], the constant A was determined and applied in equation (5) to calculate e_{eff} for the photoanode with different CdS QDs. Size. The values resulting of e_{eff} are given in Table 1. The results show that the values of e_{eff} of photoanode increased from 4785 $W.s1^{/2}.m^{2}.K^{1}$ to 6071 $W.s^{1/2}.m^{-2}.K^{-1}$ as the CdS QDs size increase from 3.21 nm to 4.4nm. The measured values of α and e ($3.2x10^{-5}$ m^2/s and 1502 $W.s^{1/2}.m^{-2}.K^{-1}$) for FTO/TiO_2 substrate are used to calculate its thermal conductivity k ($=e\sqrt{\alpha}$ = 8.5 W/m. K).The calculated values of our photoanode effective thermal conductivity (k_{eff}) of CdS QDs sizes are also given in Table 1. There is an increase of k_{eff} with increasing the CdS QDs sizes, and is about 186 % for largest size 4.4 nm

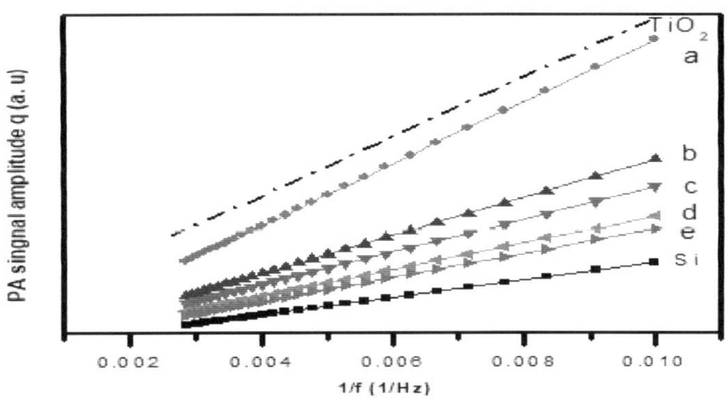

Figure 7. PA signal amplitude versus 1/f for sample a to sample e

Table 1. The calculated values of effective thermal diffusivity α_{eff}, effective thermal effusivity e_{eff} and effective thermal conductivity k_{eff} for TiO_2/ CdS QDs samples of different particle sizes.

sample	CdS particles sizes nm	(α_{eff}) (10^{-5} m^2/s)	(e_{eff}) ($Ws^{1/2}m^{-2}K^{-1}$) ±7	(k_{eff}) (W/m. K)
A	3.21	7.1±0.05	4785	40.32± 0.14
B	3.4	8.4±0.05	4975	45.60 ± 0.14
C	3.7	9.6±0.05	5383	52.75 ±0.13
D	4.01	13.1±0.05	5908	67.63 ± 0.11
E	4.4	15.4±0.05	6071	75.35± 0.11
FTO/TiO_2	-	3.2±0.05	1502	8.5± 0.12

It is easily observed that, as the number cycles increase the values of α_{eff}, e_{eff} and k_{eff} are increase. In an earlier work [15, 25], the thermophysical parameters of CdSe and CdTe QDs determined using PA technique, showed that the values of α, e and k decrease as QDs size increases. In the present case of SILAR technique, the produced CdS QDs are produced impregnated in the TiO_2 pores; decreasing the space, hence increase the thermal conductivity of the materials [26]. This results is in different to the regular chemical sensitizes technique.

3.3. Photovoltaic performance of CdS QDSSCs

The J-V characteristics curves of CdS QDSSCs are shown in Figure 8. The photovoltaic parameters V_{oc}, J_{sc}, FF and η for CdS QDSSCs are given in Table 2. .It is clearly seen that as CdS QDs size increases, the values of J_{sc} and η increase. The maximum values of J_{sc} and η are 2.24 mA/cm^2 and 0.48% for sample e 4.4 nm (10 cycles) sample e of CdS QD. S.Abdallah et. al [27,28] has made measurement on CdS QD sensitized TiO_2 using the direct adsorption (DA) technique. They reported that, an increase of J_{sc} (from 0.31 mA/cm^2 to 0.67 mA/cm^2) and η (from 0.05% to 0.18%) as the particle size increase 3.14 to . 3.84 nm. The present results shows an increase of J_{sc} from 1.18 mA/cm^2 to 2.24 mA/cm^2 and η from 0.15% to 0.48% with increasing particle size from 3.21 nm to 4.4 nm of CdS QDs compatible with previous work. with their work .They attributed the increase in J_{sc} with increase particle size in and red shift of the absorption edges harvesting more visible photons in the solar spectrum and thus causes relativity high absorption of the incident photon from solar spectrum. The Improvement of the J_{sc} and η in the case of SILAR technique is being better method than DA give high-quality surface spread, lower aggregation and higher photocurrents, [29].

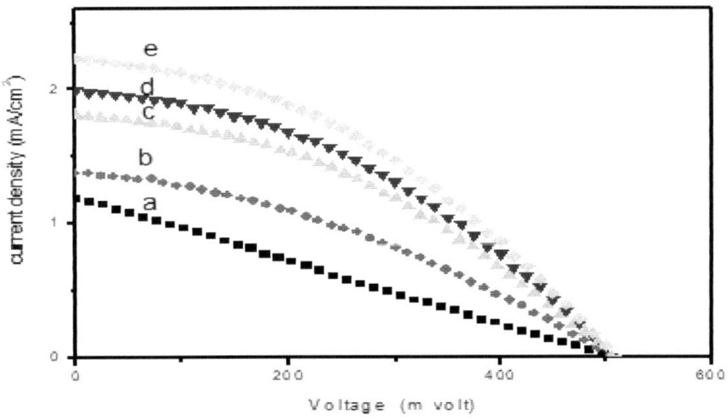

Figure 8. J-V characteristic curve of QDSSCs of CdS QDs with different different cycle

Table 2. J -V characteristics parameters of a CdS QDSSCs for different QDs sizes, under 1 sun illumination.

Sample	CdS particles sizes nm	V_{oc} (m Volt) ± 0.01	J_{sc} (mA/cm^2)	FF	η(%) ± 0.01
a	3.21	508	1.18± 0.01	0.24	0.15
b	3.4	505	1.38± 0.02	0.35	0.24
c	3.7	504	1.82± 0.01	0.39	0.37
d	4.01	509	2.01± 0.01	0.38	0.42
e	4.4	510	2.24± 0.01	0.40	0.48

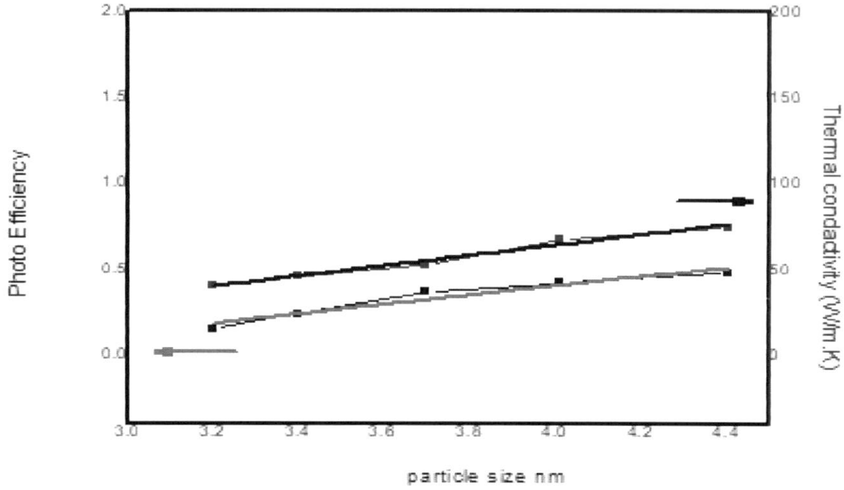

Figure 9. The relation between thermal conductivity k and photo efficiency η with QDs particle size

The variation of η and k_{eff} as function CdS QD particles size are shown in Figure 9. There is an increase in η and k_{eff} as the size of CdS QDs increase. The increase of thermal conductivity would lead to the decrease the dissipation of local heat in the solar cell and less recombination.

4. Conclusion

CdS QDs of different sizes were adsorbed onto TiO_2 nanoparticle electrode using SILAR method of varying to obtain varying layers of CdS QDs as sensitizer for QDSSCs. EDX measurements ensure the success of adsorption of CdS QDs onto the TiO_2 electrode. The UV spectra show a shifted to lower energy region from 3.07 eV to 2.70 eV as the number of cycles increase from 2 to 10 cycles. The average particle sizes were estimated using EMA to be 3.21 nm to 4.4 nm which was confirmed by TEM. The thermophysical parameter e_{eff}, α_{eff} and k_{eff} of FTO/TiO_2/ CdS photoanode of different cycles, are determined by PA technique, show an increase with increasing cycle. The increase of k_{eff} is explained by, the increase of number of cycle results a reduction of air and oxygen molecules impregnated inside the TiO_2 microspores. The increase of values of J_{sc} and η of CdS QDs solar cells with the number of cycles i is mostly attributed to better matching with solar spectrum. Furthermore, the increase of thermal conductivity leads to the decrease the dissipation of local heat and hence less recombination.

5. Acknowledgements:

The authors would like to express their sincere gratitude to Ain Shams University and the Physics Department at Ain Shams University for their financial and technical support in performing and finishing this work.

Referances

[1] Okate, K. , Kunita,S. , Jagtap, C. V. , Baviskar,P. K. , Jadkar,S. R., Pathan, H. M., & Mohite, K. C. (2018).*Optik-International J. for Light and Electron Optics*, 157 ,628-634.

[2] Tubtimtae,A. , Lee,M. W., & G. Wang,J. (2011). *J. Power Sources* ,196 ,6603–6608.

[3] Salant,A. , Shalom,M., . Hod, I, Faust,A. , Zaban A. , & Banin, , U. (2010). *ACS Nano*, 4 ,5962–5968.

[4] Kongkanand,A., Tvrdy,K., Takechi,K., Kuno,M., & Kamat,P.V. (2008). *J. Am. Chem. Soc.*, 130, 4007–4015.

[5] Badawi,A., Obdallah,W., AL-Baradi,A.M. & AL-hosiny, N. (2019). *Mater. Sci. Semicond. Process.*,95,1-12.

[6] Badawi,A., Al-Hosiny,N., Abdallah,S., & Talaat, H., ,(2013). *Mater. Sci.-Poland* ,31, 6–13.

[7] Badawi,A., Al-Hosiny,N., & Abdallah,S. (2015). *Superlat. and Micro,*. 81, 88–96.

[8] Zhang,Y, Zhu,J., Yu,X., Wei, J., Hu,L. & Dai,S. (2012). *Sol. Energy*, 86,964–971.

[9] M.P. Valkonen, S. Lindroos and M. Leskelä, *Appl. Surf. Sci.* 134 (1998). 283–291.

[10] Laukaitis,G. , Lindroos,S. , Tamulevičius,S. , Leskelä,M. , & M. Račkaitis, (2000).*Appl. Surf. Sci.* ,161,396–405.

[11] Bailey ,R.E., & Nie, S., (2003) .*J. Am. Chem. Soc.*, 125, 7100–7106.

[12] Badawi,A,. *J Mater Sci: Mater Electron,* DOI 10.1007/s10854-016-4781-1.

[13] Badawi,A. ,(2016) . *Superlatt. and Micro.,* 90 124-131.

[14] Badawi,A., Al-Hosiny, N., Abdallah, S., Merazga,A., & Talaat, H. ,(2014). *Mater. Sci. Semicond. Process.,* 26, 162.

[15] Al-Hosiny ,N., Badawi ,A., A.Moussa,M.A., El-Agmy, R., & Abdallah,S. (2012).*Int. J. Nanopart* 5,258–266.

[16] Abdallah ,S., El-Brolossy,T.A., Negm,S. , & Talaat,H. (2008).*Eur.Phys.J.Spec. Top.*,153,199–202.

[17] A.Rosencwaig,A.Gersho, (1976).J.Appl.Phys.47,65–69.

[18] El-Qahtani ,Z., Badawi , A., Easawi ,K., Al-Hosiny, N. , & Abdallah,S. (2014),*Mater. Sci. Semicond. Process.* 20,68–73.

[19] Syrrokostas,G., Giannouli,M., &Yianoulis,P. (2009), *Renewable Energy* ,34,1759–1764.

[20] Lee,Y., & Lo,L. ,(2009) *Adv. Funct. Mater.* ,19,604–609.

[21] Talaat,H.,Abdallah,T.,Mohamed,M.B.,Negm,S., &El-Sayed,M.A.,(2009),*Chem.Phy.Lett.*,473,288-292.

[22] Khalid,A., Easawi,K., El-Shaarawy,M. G., Negm,S., & Talaat, H. (2010). *Egypt. J. Solids,* 33 ,211-218.

[23] El-Brolossy,T.A., Abdallah,S., Abdallah, T., Mohamed,M.B. , Negm ,S., & Talaat,H., (2008*). Eur. Phys. J. Special Topics*, 153, 365.

[24] Raji,P. , Sanjeeviraja,C. , & Ramachandran, K., (2004). *Cryst. Res. Technol* ,39,617–622.

[25] Badawi,A., Al-Hosiny,N., Abdallah,S., Negm,S., & Talaat,H. ,(2012) .*J. of mater. Scien. and engineer.*,1,1-6.

[26] El-Brolossy, T.A., ,(2012).*Indian J. Phys.* ,86,39-44.

[27] Abdallah, S. , Al-Hosiny N., & Badawi,A., (2012)*MSTI-Nanoteck*,1,55-60.

[28] Abdallah, S. , Al-Hosiny N., & Badawi,A., (2012) *J. of Nanomaterial,*1,1-6.

[29] Li ,W., & Zhong X., (2015) *J. phys. Chem. Lett.* , 6,796-806.

IOP Conf. Series: Materials Science and Engineering **762** (2020) 012008 doi:10.1088/1757-899X/762/1/012008

Producing High Workable & High Strength Concrete Mix Having 6 Hour Retardation

Mostafa Abd Elrazek[1], Yasser Abdel Shafy Gamal[2]
[1] Lectures of properties and strength of materials, Civil Department, High Institute of Engineering Technology, EL-MINA.
[2] Lectures of structural engineering, Civil Department, High Institute of Engineering Technology, EL-MINA.

E-mail: (dr.mostafa.abd.k@mhiet.edu.eg) , (Yasser.gamal2310@mhiet.edu.eg)

Abstract. The goal of this study is to work out different experiments for retarding concrete setting time, which needed to reach slump 15 cm after six hours from mixing of concrete having grade C700. Six hours comprise 1.5 hour for transporting and 4.5 hours needed for constructability method, this long time for casting concrete is owing to a long transporting time of the concrete from the plant to the site and also the time needing for casting the concrete by using the slide form of high ring concrete wall, for these reasons there are needs to retard the concrete setting time with increasing the workability for concrete, using super-plasticizers is one of the best methods to improve the properties of the concrete, from fundamental experimental investigations on retardation, setting time and the concrete mixture with high strength of 626 kg / cm^2 after 7days and 755kg / cm^2 after 28 days can be achievable by using super-plasticizers

1. Introduction

Concrete has been developed quickly in the last decades to save a significant need for the construction industry in the world so that it considers the most common building material for many years and is expected to remain also in the future decades. And hence the concrete considers is a vital construction material that has extensive usage around the world. The concrete has distinct properties like Durability, Workability, fire resistance, and any molding shape, which lead to widespread use. Different forms of concrete have been created and built for various infrastructures like huge concrete dams, reinforced concrete buildings, concrete bridges, and precast concrete components.

To improve concrete quality, there are many factors that engineers should identify to reach the required performance of constructability, and it can be realized by strength, workability, and durability, which is related to real properties.

The quality of concrete in a structure is identified not only by the appropriate selection of its gradients and their proportions, but also by suitable methods in the casting, production, transportation, compacting, and curing of the concrete of the actual structure, often on-site. Although the actual quality of concrete can be affected by these processes [1]

The level of available design of the Mixture doesn't the same level of structural design. It is more suitable to determine as a procedure for selecting the type of mixture ingredients and their proportions. The required properties of the mixture and its proportions can be achieved by methods of a trial batch, this type of mixture is called a designed mixture.

The components of concrete have been created from the primary four (Portland cement, water, coarse and fine aggregate) to comprise chemical and mineral admixtures.

Setting can happen as water reacts with Portland cement (POC), which then gradually hardens with cohesive aggregates binding. The longer this process goes on, the less the concrete is workable enough to be placed and finished during construction.

To increase workability, it should be added water, but it leads to a shortage in the strength of the concrete and hence its durability. So that using additives can be the solution to delay the setting time of the concrete, the chemical additives, including water reducers, retarding admixtures, and hydration stabilizers, can be used for delaying the setting. Water reducers means decreasing the amount of water, which is necessary in a mixture process, creating it possible by extend the setting time without leakage in workability and durability. On another meaning, it can be slow the interaction between water and cement, and Hydration stabilizers delay the beginning of the setting.

Retarding admixture means an admixture that can be used to delay the setting time of concrete [15[15]. And it may cause cement set retardation by any of the following processes:

(1) Adsorption of the retarding compound on the surface of cement particles, forming a protective layer which slows down hydration interaction;

(2) Adsorption of the retarding includes nuclei of calcium hydroxide, poisoning their growth, which is necessary for the hydration process of cement after the end of the induction period;

(3) Formation of the congregation with calcium ions, increasing their solubility and discouraging the creation of the nuclei of calcium hydroxide; and

(4) deposition around cement particles of insoluble derivatives of the retarding components created by interaction with the highly alkaline aqueous solution, forming a protective layer [16[16]

Concrete mixtures also include the air-entraining factors, detergent-like additives that increase the durability of concrete by boost the creation of air bubbles within the concrete.

The setting of concrete can convert the fresh concrete from the liquid stage to the solid stage. It is crucial to determine the setting time to extend the duration required for transporting and placing of concrete. And it can be known as the beginning of rigidity of the fresh concrete. It is different from hardening, while it depicts the growth of identifiable strength. Setting precedes hardening, but it can be confirmed that both of setting and hardening are gradual changes, which are controlled by the cement hydration process.

On the other hand, the initial set expresses nearly the time at which fresh concrete can't be casting or handled, while the final set nearly the time at which the hardening starts. Fresh concrete will have lost measurable slump before the initial set, while measurable strength will be almost achieved after the final set.

Knowledge of the setting properties is vital in the field of concrete industry. It can be very useful in managing scheduling through the life cycle of the construction project such as concrete transporting & placing & compacting and finishing. All information about the construction method or the period of transporting is necessary to decide the use of a retarding admixture or accelerator or not, and the hydration process starts immediately when the water mix adds to the cement. The different mineral admixtures have different chemical, mineralogical compositions, and different particle characteristics, they could have various components affecting the characteristic of concrete inclusive of the concrete setting.

Slump is the most essential test for workability (e.g., ASTM C 143 or BS 1881:Part 102). The main goal of this test is to relate the change in water content with changing in the slump. For illustration, increasing the slump value leads to a lowering in the strength of concrete.

Nevertheless, various factors are affecting also in the changing of a slump not only higher water content but also the cement components as well as the grading and shape of aggregates
To increase the slump value of the concrete, it can be achieved by adding chemical admixtures or lower water content. Changes in cement component or cement fineness often lead to a change in the effectiveness of chemical admixture and hence the changing in a slump. The slump test is considering a necessary test to clarify any changing inconsistency or fluidity during the transporting of the concrete and also retarder the setting time of concrete. However, the cause of the change is not easy to detect.
Workability often refers to the ease of transporting, casting of concrete, and mixing without excessive bleeding or segregation. So that there are more needs to compare between different mixtures. [1]
No single test can survey every one of these properties (workability, durability, slumping, etc). In fact, most of these cannot be easily assessed even though some standard tests have been established to evaluate them under specific conditions (not always similar to that occurring on-site)
The primary purpose of the current study is to investigate the workability and to slump, the experimental studying helps us to increase setting time for transportation operation from the plant to a far construction project and in case of retard the setting time during the pouring of specialist construction
In recent years, many types of research have been studied in different cases about setting time, workability, and durability, which are the most critical factors that affected the characteristics of concrete.
B. Sri Umniati, Puput Risdanareni1, Fahmi Tarmizi Zulfikar had an undertaking to expand the workability of geopolymer concrete by utilizing retarder admixture (Plastocrete RT6 Plus). Using distinct percentages of fly ash quantity (0.2%, 0.4% and 0.6%) in the satisfied of retarder. To perform the direct, geopolymer concrete without retarder (0%) were also made. The activator manner in this study was Na_2SiO_3 add to NaOH10M discharge, with a proportion of 1:5. The terminate liable an optimal makeup of geopolymer concrete with 0.6% retarder, where initial setting time occurs after 6.75 hours, the final setting time occurred after 9.5 hours. furthermore, the geopolymer concrete slump was 8.8 cm, and the slump flow was 24 cm. [2]
The hot weather leads to the fast evaporation of water from the surface of the fresh cement paste. Using the retarders to exchange the accelerating realization of hot weather on the concrete setting time, and it can be treating the effects of various extraordinary conditions. [3]
The retarder mixture can be used in retarding the setting time of concrete [4]This phenomenon can lead to one or more of the process among which; the admixture molecules are adsorbed on the surface of the cement particles, and it can be stopping the reactions between cement and water.
M. H.R. Khudhair, M.S. Elyoubi [5]have reached to methodology to enhance the mechanical performance of mortars and/or concrete, including the porosity /the capillary absorption / the compressive strength and decreasing the amount of mixing water used.
The chemical additives can affect the setting and hardening of Portland cement. Different types of retarders and accelerators are known. the commitment recommendations from the American Concrete Institute clarify the narrative of the effects of additives upon the characteristics of concrete [6]

In the past, the reaction of the organic material in the retarding setting is not known and not fully understanding the effects of their reactions [7] Taplin [8] supposed that organic substances retard by adsorption either on the surface of the clinker minerals to protect them from water attacks, or on the surface of a coherent coating of hydration products to stop the transporting of material to or from the clinker surface.

Mortars containing ordinary Portland cement and micro-silica setting time are more significant than mortars with nano-SiO_2 particles setting time. [9]

Bazid Khan and Muhammad Ullah [10] discovered that the effects of admixture on setting time of cement pastes are relying on the type of cement and dose of the admixture. It can be lead to set retardation of the three different kinds of cement used,

Cement Paste was created, which gradually stiffens and hence hardens. Setting means the stiffening of cement paste [11]

Fundamentally, the setting is a mechanism to change the concrete from an initial state, a scattered concentrated suspension, to a final state, a connected and strengthened system of particles. The chemical reactions between cement particles and water (cement hydration) can cause this transformation in the practice of cement and concrete [12]. Normal setting time of cement is referred to as hydration of Alite (impure C_3S) and composition of the calcium silicate hydrate (CSH) phase [13]

Using the retarders to inhabit the accelerating effects of high temperatures which reduced the setting time, or to solve the problems results from unfavorable delays between mixing and placing occur [14]
Concrete setting properties are the most vital part of the concrete construction phase (Brooks et al., 2000) [17] It assists in the evaluation of different concreting processes transporting, casting, consolidation, and hydration. The casting of fresh concrete in the formwork relies on the concrete setting time, which makes the concrete hardening (Clear and Harrison, 1985) [18]
The hard behavior of the matrix can specify the final and initial hardening time of concrete. The initial setting time of the concrete points to the onset of the hardening of the mixture and the final setting time points to the adequate hardness of the concrete mixture (Naik et al., 2001) [19]
(Brooks et al., 2000) Studies showed that increasing of content of fly ash within the binder follows with increasing in setting time [17] Carette and Malhotra, 1984) [20]
Niragi Dave a, Anil Kumar Misra , Amit Srivastava, S.K. Kaushik [21]found that setting time of tripartite blended concrete created from FA and GGBs referred to retarding initial hardening time in the range of 60–120 min, in the thesis using additives to achieve the set time after 6.5 hours is our target to save construction requirements
M. Davraz [22]discusses the measurements of Setting times dependent on PH, EC, and T changes of fresh mortars. With increasing BA concentration, ph, electrical conductivity, T values of the mortars will reduce, and on the other hand, setting times of mortars will increase.
Chaiyakrit, Ram, Chalermchai [23]solve the problem of the slump of concrete by using the second dosage of superplasticizer (naphthalene based type F) which add during the process to recover the slump loss, eventually, the compressive strength and setting time of concrete can achieve after dosage with chemical admixtures type F (superplasticizer, naphthalene based) for slump recovery. On the other hand, the addition of the second dose of superplasticizer will rise, setting times accordingly to additives dose. As for fly ash concrete, the prolongation effect was higher as the replacement of fly ash increase. The prolongation effect can reach up to 4 hours. The acceptable range of compressive strength for the re-dosed concrete ranges between ±10%
Megahd Ahmed H., M. M. Rashwan , and Mostafa A. showed that After 25 cycles of durability tests, control specimens with the control admixture (secrete wp) and without admixture showed a considerable decrease of their compressive strength. But specimens modified with the suggested admixture (BM 2010) showed respectively a small increase of their compressive strength by about 7% if compared with the same specimens before these cycles of durability test. The suggested admixtures showed a better influence compared to the control admixture secrete wp [24][25]

2. Experimental

The characteristic of materials was used in the experimental to create concrete mixes as followings:

2.1 Materials

2.1.1.Aggregate

The aggregate used in the experimental crushed basalt (as a coarse aggregate), sand (as fine aggregate) have properties as in table (1), (2), and Fig. (1 a&b)

Table (1): Properties of Aggregate

Property	Basalt	Sand
Volume weight (t/m3)	1.59	1.68
Specific gravity	5	2.5

Table (2): sieve analysis of aggregate.

Sieve size	% Passing by weight.	
	Basalt	Sand
40mm	100	---
20mm	98.31	---
10mm	39.94	---
5mm	1.97	---
2.5mm	---	100
1.25 mm	---	99.96
0.63 mm	---	88.57
0.31 mm	---	58.35
0.16 mm	---	18.7
No. 100	---	8.13

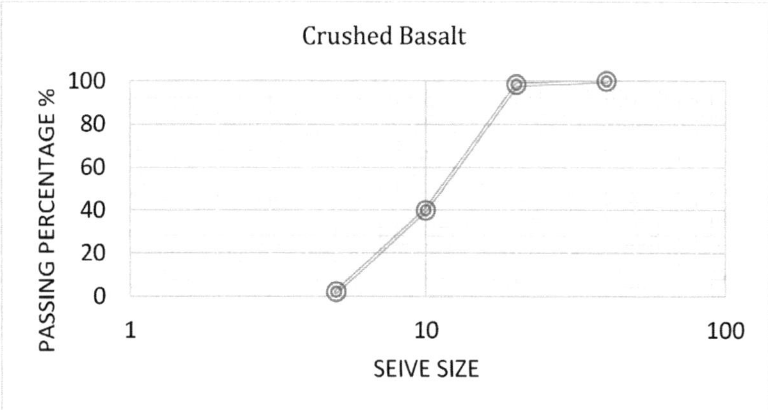

Fig. 1a. Aggregate sieve analysis (crushed stone)

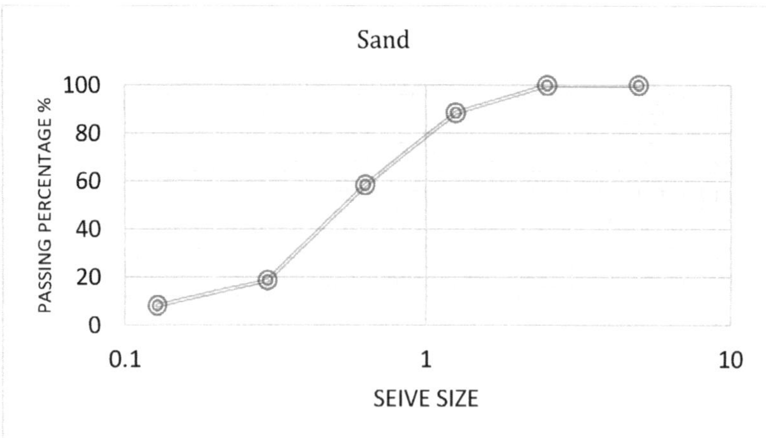

Fig. 1b. Aggregate sieve analysis (sand)

IOP Conf. Series: Materials Science and Engineering **762** (2020) 012008 doi:10.1088/1757-899X/762/1/012008

2.1.2. *Cement*

Ordinary Portland Cement (OPC) conforming to Egyptian standard was used. The physical properties of the used cement are shown in Table (3)

Table (3): Cement Properties.

Fineness of cement		3300 cm2/gram.	3300 cm2/gram
Strength of cement	2 day	248 Kg/cm^2	≥100 Kg/cm^2
	28 day	425 Kg/cm^2	≥325 Kg/cm^2
Setting time	Initial	2:12 hour	≥75 minute
	Final	4:45 hour	

2.1.3. Silica fume

Silica fume was used in the process to produce silicon metal and ferrosilicon alloys, Properties of silica fume comprises high content of amorphous Sio2 & particle size of 0.1 -0.2 micron with spherical shape, the percentages of chemical component were taken in the experimental as shown in table (4), and the originate of production for the component is Egyptian Ferroalloys Co. in Edfo city.

Table (4): Chemical component of used silica fume.

The component	Sio2	Fe2O3	Al2O3	MgO	Na2O	K2O	Ca2O	H2O	L.O.L	C
Percentage of component	Min. 92%	Max. 1.5%	Max. 1.5%	Max. 1.0%	Max. 0.8%	Max. 1.3%	Max. 0.5%	Max. 1.0%	Max. 2.0%	Max. 1.5%

2.1.4. Water

The water used in the concrete mixtures is validated for drinking

2.1.5. Additives

Sikament NN :

A superplasticizer without retarder (ASTM C494 Type F) "A highly effective dual-action liquid super-plasticizer to produce free-flowing concrete or as a substantial water-reducing agent for promoting high early and ultimate strengths.

Sikament R2008:

A superplasticizer (ASTM C494 Type G) "A highly effective superplasticizer with a set retarding effect for producing free-flowing concrete in hot climates. Also, a substantial water reducing agent for promoting high early and ultimate strengths". with a density of 1.185 Kg/L

2.2 Concrete mixtures

The test for studying the setting time and strength concrete is based on producing high strength concrete after 28 days, the high strength concrete after 28 days were about 700 Kg/cm^2. Concrete mix trails proportion is shown in table (5).

Table (5): Concrete mixes proportions by weight

Mix No.	Cement content (Kg/m^3)	Sand content (Kg/m^3)	Coarse Agg. Content (Kg/m^3)	Silica Fume content (Kg/m^3)	Sikament R2008 Liter/m^3)	Sikament NN Liter/m^3)	Water content (Liter/m^3)	Initial Slump (cm)
1	500	727	890	50	8	8	200	25
2	500	727	890	50	8.75	8.75	200	26
3	500	727	890	50	10.75	10.75	200	28
4	500	727	890	50	11.75	8.75	200	27

A superplasticizer was used (Sikament R2008, Sikament NN). Six standard cubes 15 x 15 x 15 cm were cast from each mix and cured in laboratory

2.3 Fabrication of tested samples

the mixture of 0.1 m3 capacity can be used in the composition of the constituent materials .in the beginning, mix the dry aggregates, cement, silica fume and aggregate for a one minute and then add the water which mixed with admixtures to create continuity mixing of the component until reaching to the homogenous constituent, this process will take two minutes to finish

Erect the experimental specimens by using steel forms, before pouring the specimens, and there are some steps to follow:

- Grease all the interior form to simplify the extraction of the concrete beams
- Transporting the concrete of the plant by hand shovels with steel form
- Execution the compaction by using the mechanical vibrator
- No of specimens is three cubes with 15 cm side length
- Removing the forms after one day from pouring for all specimens.
- In the final process, the concrete specimens by submerging in the water tank until 28 days.

2.4 Instrumentation (Description of the testing machine)

Using the testing machine (MATEST to 100 tons up) to execute the static load, which loaded from zero until failure.

3. Results and Discussions

The results from lab tests done for the concrete compressive strength and the workability of the fresh concrete, just a slump test has been done to get the fresh concrete properties.

The average values of cube strength of 7 and 28 days for each mix are shown in table (6) and fig (2).

Table (6) The average of cubic strength

Mix No.	Compressive strength Kg/cm^2 (Average of three cubes)	
	After 7 days	After 28 days
1	610	668
2	616	678
3	640	711
4	626	755

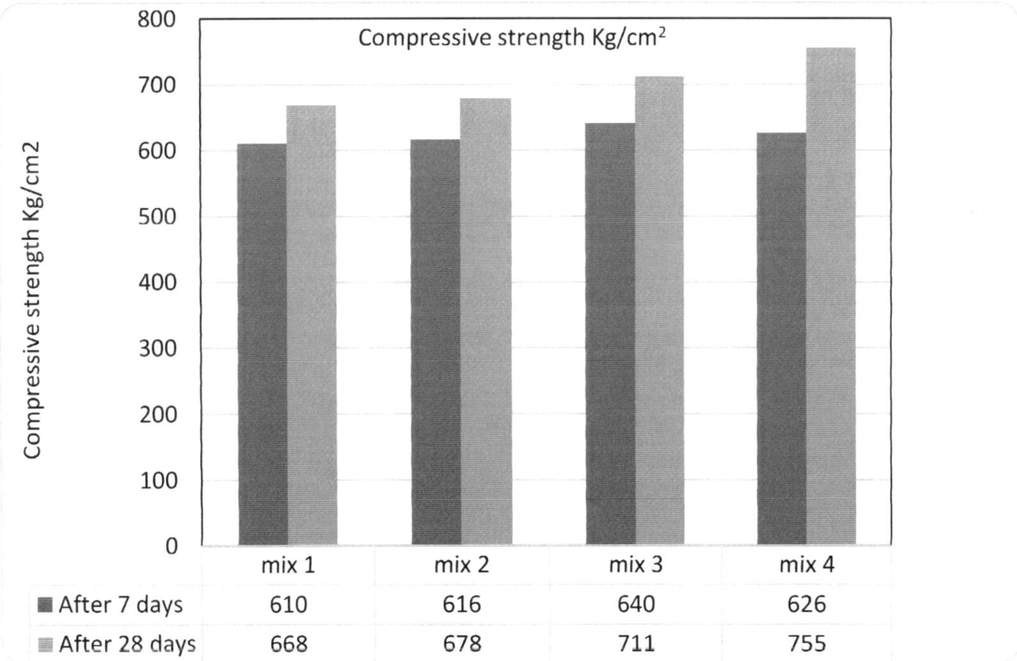

Fig.2 Compression strength of concrete mixes

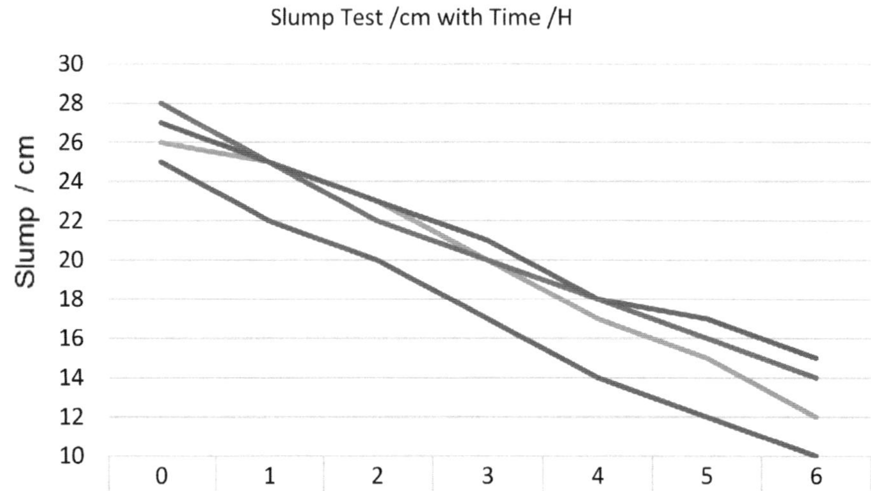

	0	1	2	3	4	5	6
mix 1	25	22	20	17	14	12	10
mix 2	26	25	23	20	17	15	12
mix 3	28	25	22	20	18	16	14
mix 4	27	25	23	21	18	17	15

Fig. 3 workability loss based on slump with time

Effect of mixing two types of superplasticizer with retarder and superplasticizer without retarder on Slump Loss has been investigated. The results of slump loss for superplasticizer with and without retarded on concrete shown in Fig 3 and 4.

Figures confirmed that slump decreases with increasing time. It also showed that the hydration technique produces calcium silicate hydrate, which fills the pores between the combination and particles of aggregates and cement. As a result, the setting of the concrete will reduce concrete fluidness, hence, reduce the slump too. When observation is carried out on adding of super plasticized with and without retarded, growth in dosage of the chemical admixtures will increase concrete setting time, in view that both of the superplasticizers with and without retarder will help to preserve the concrete liquidity for an extended time, and as a result, lessen the slump loss during transportation of concrete to site. However, the overdosage of these admixtures will not cause excessive slump. If comparisons are made between the superplasticizers with and without retarder, the sitting time for retarded concrete is longer than the plasticized without concrete. After 360 mins (6 hours) as a result, the end showed that retarder gives higher power in keeping the concrete slump than the superplasticizers without retarder does at the same dosage.

From results, the increase of the dose of superplasticizer with retarder to 2.35% mixed with superplasticizer without retarder dose 1.75% will gain the minimum slump loss with time needed (6 Hours).

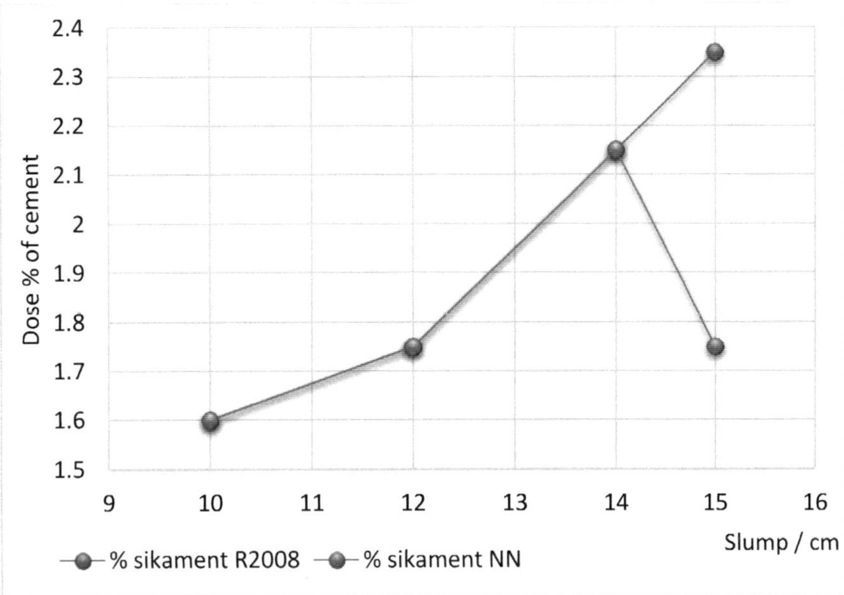

Fig. 4 Effect of super plasticizers with and without retarder on workability loss based on slump

The effect of adding of superplasticizers with and without retarder on Compressive Strength of concrete with different dosages are shown in Fig. 5 and 6. This test is performed at age 7 and 28 days. The values of compressive strength for the different dosage of superplasticizer with and without retarder are then shown as in Fig 4 and 5 & 6. From the figures, strength gains from adding chemical admixture is discovered by way of the growth in compressive strength with age. At an early age (7 days from casting), the strength is high for a reason that reaction of cement particles with water is active. Hence, the slope of the curve for age 28 days is less steep compared with an early age.

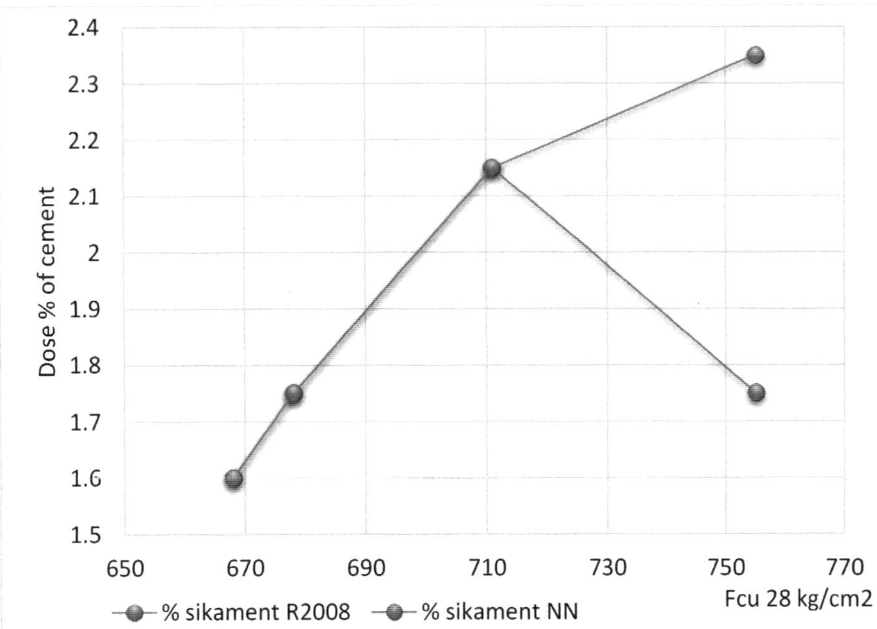

Fig.5. Effect of superplasticizers type and dosage on compressive strength After 28 days

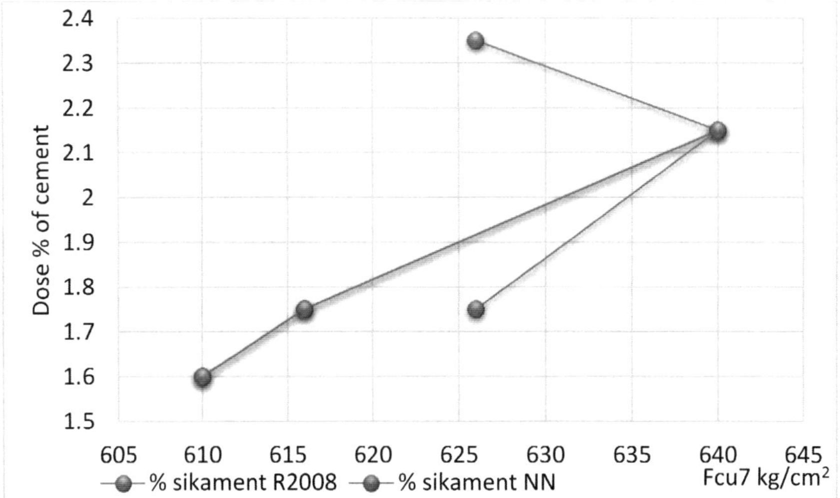

Fig.6. Effect of superplasticizers type and dosage on compressive strength after 7 days

Mix of admixtures doses show different behavior on the cubic strength of concrete. At age 7 days, the growth of dose of superplasticizer with retarder growth of concrete cubic strength until dose 2.15%, on the other hand, more increasing of dose until 2.35% reduces the strength, the cause for this phenomenon is that addition of retarder to the concrete will put off the reaction of C3S and C3A. As a result, strength development is low. This varies happened only after 7 days from casting for adding the same dose from each admixture type, in which inclusion of superplasticizer with retarder will improve cubic strength at a later age. For superplasticizer without retarder, the increase of dosage will increase the cubic strength for every age also until dose 2.15%. Considering that the addition of superplasticizer without retarder will provide water for concrete mixing, the hydration technique will not be disturbed, but it is expanded with adding of additional water from the deflocculation of cement particles. Hence, growth in dosage will increase the entrapped water and promote hydration of cement. Though the growing dosage of admixture will increase the compressive strength, there will be nonetheless a top of the line limit for using admixture. While the dosages go beyond this limit, the boom in dosage will become most effective in reducing the compressive strength.

This phenomenon occurs because the overdosage of superplasticizers with retarder & superplasticizers outstanding of retarder will purpose bleeding and segregation, to be able to affect the cohesiveness and uniformity of the concrete. As a result, cubic strength will less if the used dosage is beyond the optimum dosage. For early age, cubic strength of concrete containing a mix of plasticizers with retarder 2.15% and awesome plasticizers without retarder 2.15% exceeded to 640 kg/cm^2. While, overall performance of higher dosage decreases the cubic strength. But, for a long time (28 days), cubic strength, blend of plasticizers with and without retarder provide proper result, in which the cubic strength exceeds to 755 kg/cm^2.

4. Conclusion

This paper was conducted to study the effects of superplasticizer and retarder on properties of high strength concrete with characteristic strength of 700 kg/cm^2, which having slump 15 cm after 6-hour retardation. The properties investigated were workability (slump), and compressive strength. However, the conclusion, which follow are drawn based on experimental results and observations presented earlier. These conclusions are of necessity specific to this study, is related to the type of superplasticizer with retarder and superplasticizer without retarder, an environmental condition during testing, testing method, etc. Nevertheless, the findings of this investigation should provide a significant contribution towards the knowledge on the effect of mixing of superplasticizer with retarder (Type G) and superplasticizer without retarder (Type F) on properties of concrete.

The properties of concrete containing superplasticizer with and without retarder (Type G & Type F) had been successfully studied. From the results of the study presented earlier, the following conclusions are offered:

- The workability of concrete can be increased by the addition of a mix of superplasticizer with retarder (type G) with dose 2.35% and superplasticizer without retarder (type F) with dose 1.75 % by weight of cement content for each one.
- Slump loss can be reduced by using the chemical admixtures superplasticizer with retarder (type G) with a dose of 2.35% mixed with superplasticizer without retarder (type F) with a dose of 1.75 %.
- Compressive strength is improved by superplasticizer with and without retarder for ages 28 days.
- Saving concrete workability with slump 15 cm until 6 hours with gaining compressive strength 755 Kg/cm^2 can be achieved by adding a mix of superplasticizer with retarder (type G) with dose 2.35 % and superplasticizer without retarder (type F) dose 1.75 % of cement content for each one (such as mix 4).

References

[1] W.f. CHEN J. y. Richard Lie W 2013 *The civil Engineering Handbook* second edition.
[2] B. Sri Umniati, Puput Risdanareni, Fahmi Tarmizi Zulfikar Zein 2017 *AIP Conference Proceedings* "Workability Enhancement of Geopolymer Concrete through The Use of Retarder" 1887, 020033 ; https://doi.org/10.1063/1.5003516
[3] H. G. Ni, J. Z. Wang, Cem. Concr. Res, (2000) **30.8** 1245-1250.
[4] A. Ul-Hamid, L. M. Al-Hems, A. Quddus, A. I. Muhammed, H. Saricimen, Anti-Corros. 2017 *Methods Mater* **64.3**
[5] M. H.R. Khudhair, M. S. Elyoubi, A. Elharfi 2018 *J. Mater. Environ. Sci* Study of the influence of water-reducing and setting retarder admixtures of polycarboxylate "superplasticizers" on physical and mechanical properties of mortar and concrete, **9.1**, 56-65
[6] Foster, B. E. (ed.). 1963 *Admixtures for Concrete* (Third Report of ACI Committee No. 212). Jour., ACI (Proc.), **60**, 1481
[7] Vivian, H. E. 1960 *Fourth Internat. Symposium on Chem. of Cement* Some Chemical Additions and Admixtures in Cement Paste and Concrete., Washington. Proc., p. 909.
[8] Taplin, J. H., 1960 *Discussion. Fourth Internat. Symposium on Chem. of Cement*, Washington. Proc., p. 924
[9] Afshar Ali Hosseini, Seyyed Hossein Hosseini, and Ali Reza Abbas Zadeh 2012 *Asian Journal of Chemistry* "Study of Compress Strength and Time Setting of Concrete by Additives of Silica Fume and Nano Silica" **24.2** 903-907.
[10] Bazid Khan* and Muhammad Ullah 2004 *JKAU: Eng. Sci.* "Effect of a retarding admixture on the setting time of cement pastes in hot weather" **15.1**, 63-79 (1425 A.H./ A.D.)
[11] Baradan, B. (1998). *Construction Materials* –II (5th.ED). Dokuz Eylul University, Technical faculty publication section, Izmir Turkey.
[12] Jiang S. P., Mutin J. C. & Nonat A. (1995). *Cement & Concrete Research* Studies on mechanism and physio-chemical parameters at the origin of cement setting **25.4** 779 – 789.
[13] Chen Y. & Older I. (1992). *Cement & Concrete Research* On the origin of the Portland cement setting. **22.6** 1130 – 1140.
[14] United States Department of Transportation – Federal Highway Administration (1999). www. Fhwa.dot.gov/engineering/hng20/hng23/setretrd.htm
[15] Annual Book of ASTM Standards (1982), part 14, concrete, and mineral aggregates. C 494-81: standard specifications for chemical admixtures for concrete.
[16] Banfil, P.F.G., & Saunders, D.C. (1986). *Cement and Concrete Research* The relationship between the sorption of organic compounds on cement and the retardation of hydration., **6.3** 399 – 410.

[17] Brooks, J.J., Johari, M.A., Mazloom, M., 2000, *Cement Concr. Compos.* Effect of admixtures on the setting time of high strength concrete. **22.4** 293–301
[18] Clear, C.A., Harrison, T. A. 1985. *Concrete pressure on formwork*, in CIRIA report R 108, London.
[19] Naik, T.R., Singh, S.S., Ramme, B.W., 2001. *in The 7th International Conference on Fly Ash, Silica Fume, Slag, and Natural Pozzolans in Concrete (CANMET/ACI): Proceedings* Time of setting influenced by inclusion of fly ash and chemical admixtures **1**. Ed. By V.M. Malhotra. July 22–27, 2001, Madras, India, 393–413.
[20] Carette, G.G., Malhotra, V.M., 1984. *In CANMET Technical Report MRP/MSL* Characterization of Canadian fly ashes and their performance in concrete. 8U–137, Ottawa.
[21] Niragi Dave a, Anil Kumar Misra, Amit Srivastava, S.K. Kaushik 2017 *International Journal of Sustainable Built Environment "Setting time and standard consistency of quaternary binders: The influence of cementitious material addition and mixing"* **6.1** 30-36.
[22] M. Davraz 2015 "The Effect of Boron Compound to Cement Hydration and Controllability of this Effect " DOI: 10.12693/APhysPolA.128.B-26. Vol. 128) **128**
[23] Chaiyakrit Raoupatham, Ram Hari Dhakal, Chalermchai Wanichlamlert 2015 *Civil and Environmental Engineering* "Study on Compressive Strength and Setting Times of Fly Ash Concrete after Slump Recovery Using Superplasticizer" **9.12**, waset.org/Publication/10003192
[24] Ahmed, A. Megahd, M. M. Rashwan, and Mostafa A. Razek. 2013 *Journal of Engineering Sciences, Faculty of Engineering, Assiut University* " Mechanism effect of an advanced local and economical additive on the compressive strength and durability of concrete" **41.1** 29-47.
[25] Ahmed, A. Megahd, M. M. Rashwan, and Mostafa A. Razek. 2013 *Journal of Engineering Sciences, Faculty of Engineering, Assiut University* "Performance of a Local Proposed Economical Additive on the Cement Mortar Properties Affected by Aggressive Environmental Conditions." **41.2** 421-440.

IOP Conf. Series: Materials Science and Engineering **762** (2020) 012009 doi:10.1088/1757-899X/762/1/012009

Using the fault tree for analysis the schedule delay factors

Yasser Abdel Shafy Gamal[1], Mostafa Abd Elrazek[2]
[1] High Institute of Engineering Technology, EL-MINA Lecture of structural engineering, Civil Department, El Mina
[2]High Institute of Engineering Technology, EL-MINA. Lecture of properties and strength of materials, Civil Department

E-mail:Yasser.gamal2310@mhiet.edu.eg , dr.mostafa.abd.k@mhiet.edu.eg

Abstract. The delaying in the construction projects is one of the magnifying problems facing the construction industry in Egypt , and still a lot of researchers interest to find the main reasons for these problems and suggest the solution for mitigate or avoid the problem , a lot of contractors in Egypt suffer from the liquated damages due to the delaying problems , although there are a lot of researchers find out the delaying factors problems , but we need to get the root cause analysis of the delaying factors , in our research , the investigation find out the most critical delaying factors and categorized the factors in modeling by using factor analysis methodology into 7 factors modelling and hence using the root cause analysis to link the reasons with each other , Fault tree is one of a valuable tool can be used to link with the reasons of delaying which will be used in our research and help us to investigate deeply with the roots of the problems which is our target in the research and in the final the major problems leads to most of the problems are the delaying of the supplies (Code of Factor DF5) with mean 4.2039 and RII = 0.84078 , A lot of change Orders (code of Factor DF 14) with mean 4.1845 and RII =0.8369 , no of delaying factors in our research are 22 factors which considering as an important delaying factors in the construction industry in Egypt , the survey pilot contains no of 103 questionnaires to rank the delaying factors by statistical analysis.

1. Introduction

In the construction industry, the completion of the project on the specified time is considered as one of the important constraint for succeeding the project, not only the budget, quality but also the time, avoiding the penalty and liquated damages by ending of the project on the specified time can't be achieved without investigation on the reasons of delaying, Chan and Kumaraswamy 1997 considered the completion of the project on contract time is a signal for efficiency, a lot of features in the construction industry had been investigated such as extension time on the project, environmental conditions, the conflict in interesting between the stakeholders and the dynamic effect of the project team, there are some of the features created the obstacles in completion the project on specified time on the contract. The delaying in the construction industry is an inevitable problem facing almost construction projects,

Different researchers had been investigated about the reasons, and the researching had been happened by the survey pilot method to analysis the opinion of the questionnaires about the reasons of the delaying, understanding the reasons of problems is the root of solution, a little research interest about analyzing the root cause of the delaying so that it considers as an essential tool will be used in the research to link between different sources of delaying, the personal experience contributes in determination the importance of the delaying factors which is depends on the participants, delaying of the construction projects deduced from series of events, the events can be divided into minor and major events according to its impacts on the scheduling, the objective of research to determine the leading cause of delaying by using fault tree diagram as a tool for determining the causes of delaying which assists to link between different cause, fault tree analysis can be used to determine the delay cause and identify the relationship between the causes.

The succeeding of the construction project depends on ending the project on time within cost and with satisfaction of the stakeholders , the main problem of almost projects is delaying the project and over time than the time of the project according to the contract , indeed the time of completion of the project before or after depends on the uncertainties of risk [1] , Assaf, S.A. and Al-Hejji, S [2] deduced that about 70% of constructions projects suffer from delaying and nearly 76 % of the contractors and 56% of consultants have deduced that the percentage of increasing of time in the construction projects is in range between 10 to 30% from the original duration which lead to increasing the cost by 50% , about 50 % of the construction projects in the United Arab Emirates suffers from the delaying [3] , the delaying of the construction project is considered as the major problem caused the negative impact on the construction industry [4] , indeed the delaying of construction projects which means the time beyond the contract durations can considered as costly situation , in another meaning the delaying of the construction project interprets as the extension of time to end the scope of works in the contract [5] , Different factors can effects on the delaying of the project and a lot of researchers interest to clarify the problems which differ from country to another [6-7-8] , poor site management can effect on the extension of the duration of the project [9] , the cause of delaying in US (United State) has been investigated by Bordoli and Baldwin [10] and they clarified the climate , the labor are the main factors in delaying and the subcontractors were demonstrated the leading causes of delaying are poor risk management ,unforeseen condition, slow making of decision.

2. Literature Review:

The following the duration of project in the schedule plan is sign of succeeding the project, the completion of the project on specified time in the contract means successful project, using different tools in the project management to control the construction project duration, the professional-managerial using the management tools to achieve the objective of the project, the completion of the project on the specified time is one of the objective of the project, in the project management, using the resources (labor, equipment, materials) and methodology [11] to manage the project and end in specific time, due to the absence of the active management the delaying of the project and cost overruns have been happen.

There are different methods can be used to determine the causes of the delaying and to prevent the occurring of delaying, Using the root cause analysis as tool to achieve the previous objectives, Using the root cause analysis in discovering the cause of delaying in the construction projects, it is more effective tool to figure out the problem and solving their, and it contributes to finding out the primary causes of problem, the fault tree analysis is a kind of root cause analysis methods, it can be used to visualize the delaying factor and their relationships between causes, the fault tree analysis directs to identify the primary cause of delaying that lead to most of delaying events, RCA (Root Cause Analysis) used in the analysis of reasons of failure in the construction project [14].

The analysis of Fault tree can be assisted to help for understanding the reasons for the delaying to compose procedures to mitigate or avoid the delaying.

Fault Tree Analysis is a diagrammatic tool can be used based on the probability of occurrence or non-occurrence of the event [15&16],

Fault tree analysis is an appropriate tool to use for the delay factor analysis, especially to link the causes to each other. To execute the Fault tree analysis, it can be done in steps, in the beginning, determine the main event at the top, in our research the top problem is the delaying of the construction projects, it can be considered as a critical problem, and in the second step link the top event to other events by the logic gate (OR, And), the process can be created until all the basic cause of delaying in the construction projects, Fault Tree analysis can be qualitative or quantitative or both, the type of analysis depends on the available information, the big objective of using FTA in the delay analysis is to enable to get the prioritization causes for delaying in the construction projects.

3. The collection of data

In the starting of the research, the gathering of the data is one of the crucial steps to get the input for the research so that there are two steps to get the data of the delaying factors, the first step: interviews with various project managers in random construction projects and selecting different delaying factors, the second step: omitted the similarity delaying factors from the survey and focus only the different causes to avoid the similarity in the factors which can occur from the interview of the project manager, in the final no of the delaying factors in our research become 22 factors as shown in Table (1)

Table (1)

Ser	Item	Code
1	Delaying of issue the license and agreement to construct the project	DF 1
2	Lack of communication between different department inside the organization	DF 2
3	Effects of absence of work authorization system	DF 3
4	Absence of the communication requirement inside the organization	DF 4
5	The delaying in the procurement process	DF 5
6	The delaying of the supplies	DF 6
7	the delaying of payment request for vendors and subcontractors	DF 7
8	The delaying of the fee of the labor and incentives	DF 8
9	Absence of justice in distribution the fees between the labors	DF 9
10	Poor performance management of the subcontractors	DF 10
11	Slow making decision	DF 11
12	Hesitated and change in the decision making	DF 12
13	careless the risk management factors in the projects	DF 13
14	A lot of change Orders	DF 14
15	Lack of Skilled workers	DF 15
16	Absence of Inflation and the change in the prices of the materials, equipment and wages	DF 16
17	the poor management	DF 17
18	the weakness of the technical performance	DF 18
19	the absence of communication outside the organization	DF 19
20	Delaying of the deliverables at the working stages	DF 20
21	the effects of type of leadership on the time and cost of the project	DF 21
22	The effects of type of organization (centralized , decentralized)	DF 22

The questionnaires collected from different types of construction to get credibility and confidence in the results of analysis, there are around 103 questionnaires from construction projects with diversity in the construction types, Fig 1 indicates the types of construction work in the projects, these projects comprise 61.17 % residential projects, 7.77% infrastructure projects, 2.91 % Electromechanical construction projects, 28.16% roads projects

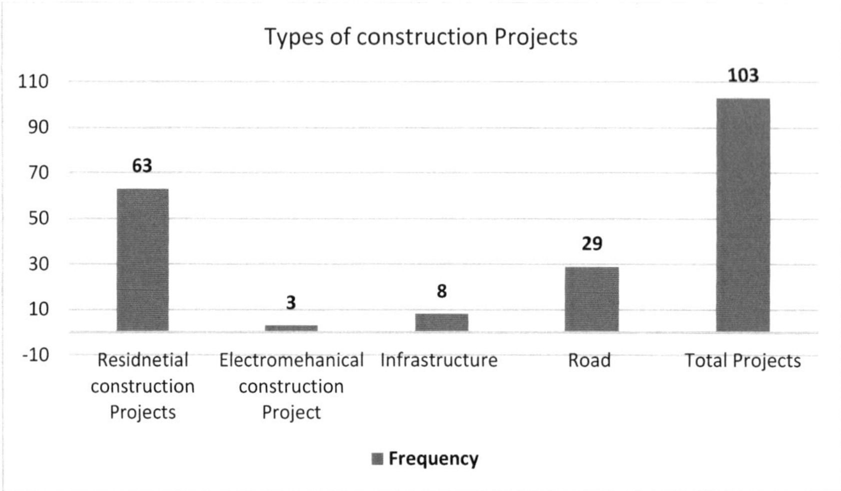

Fig (1)

The degree of education of the respondents are diverse from bachelor's degree to a doctoral degree. Fig 2 illustrates the degree of education for the respondents, the respondent 's degrees are bachelor's degree (82.52%), (12.62%) the Master 's degree and 4.85% Doctoral Degree.

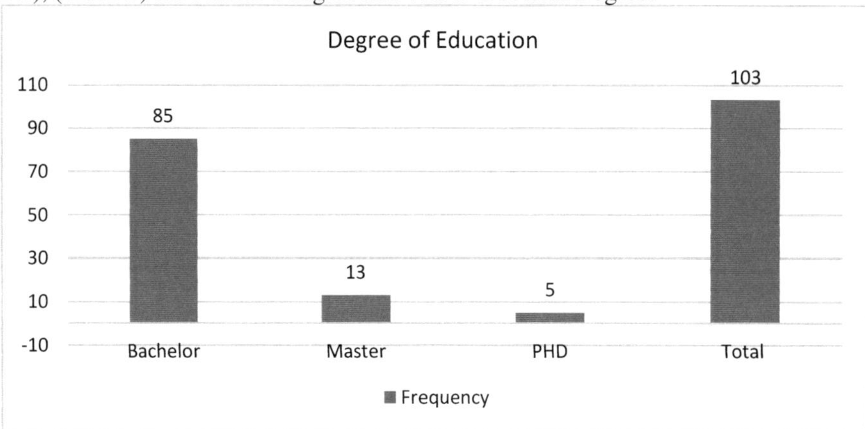

Fig (2)

The no of experience years for the respondents is illustrated in Fig (3) which clarify that no of respondents with experience between 10-20 years is 93 responders with percentage 90.29%, the experience between 21-30 are 5 responders with percentage 4.85% and in the final no of responders with experience more than 40 years are 2 with experience 2.91%

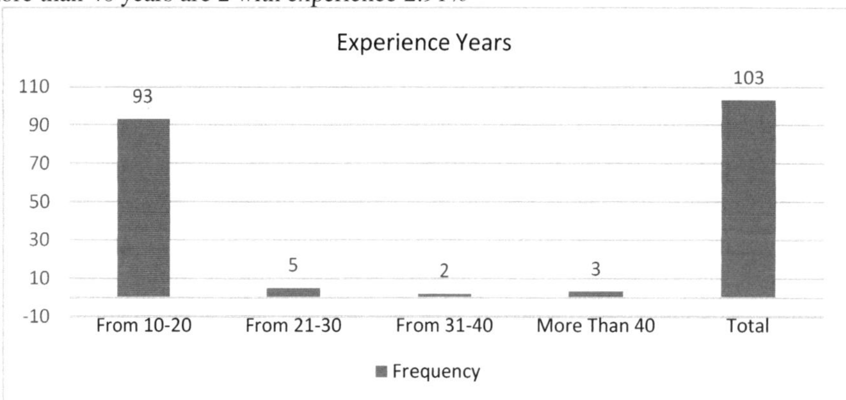

Fig (3)

The age as a factor effects on the opinion of the respondents. Fig (4) illustrates the age of respondents which indicate that the difference between the age of respondents, as shown in Fig (4), it can illustrate that no of respondents between 23 Years to 40 Years are 93 responders, from 41 Years to 50 Years are 3 responders. from 51 Years to 60 Years are 3 responders and more than 60 Years are 4 responders

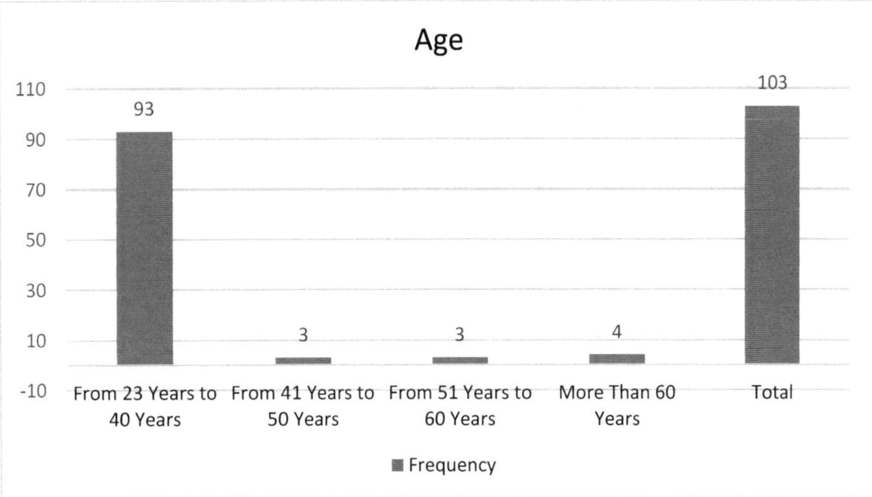

Fig (4)

In the other side, the type of respondents are 93.2% male and 6.8% female as shown in Fig (5).

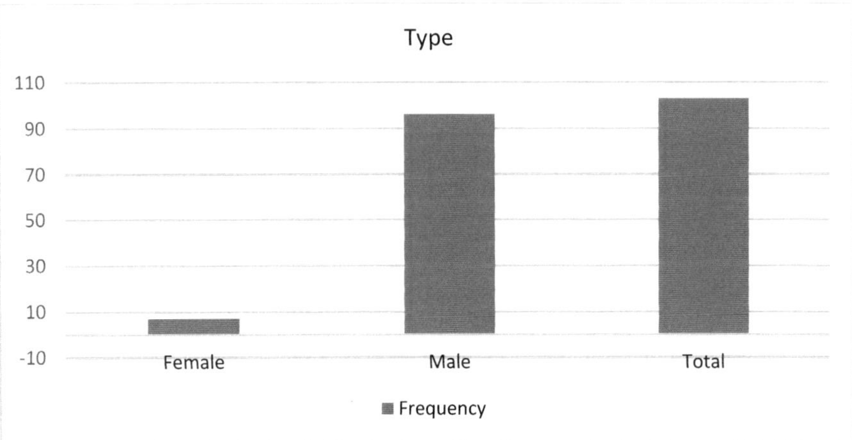

Fig (5)

4. objectives of the Research

The study has an objective to get the most impact factor effects on the delaying of the project and understanding the relationship between the causes of delayed by using a fault tree as tools for root cause analysis and to achieve the objective there are different steps have been followed to achieve the target of research and also create modeling for the delaying factors by using the factor analysis, and these can assist in managing the prevention the reasons of delaying in the construction projects.

5. Data analysis and results

The methodology of research can be illustrated in the flow chart which indicated in Fig (6), in the flow chart, it can be obviously the road map of research to get the objectives, in the beginning, First step: starting with identification of the study case (Delay of the construction project), second step: collection of data from different responders, third step : omitted the similarity delaying factors from the survey pilot, fourth step: Questionnaires survey. Fifth step: categorized the delay factors: sixth step: Analysis the delay factor by using relative importance index seven steps: using Fault Tree analysis as tools of root cause analysis .

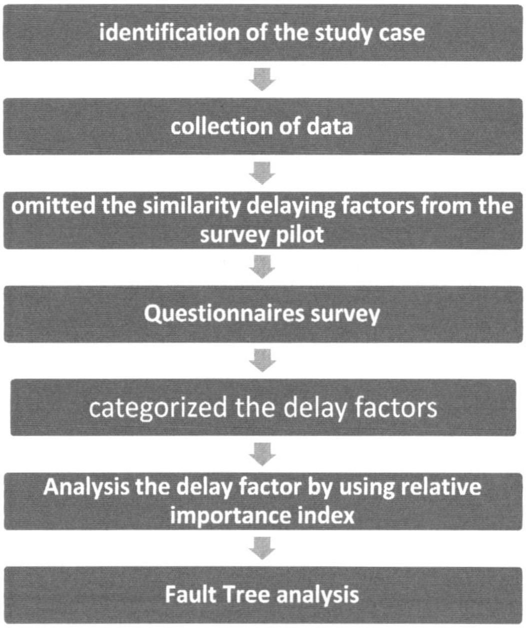

Fig (6) Flow chart of methodology of research

5.1 Alpha Cronbach

To get the reliability and internal consistency, using Cronbach's alpha as the suitable statistical method to evaluate the reliability and confidence of the questionnaires [12], the range of acceptance degree of credibility of alpha Cronbach are above 0.6 [13], in the research by calculation submitted from using SPSS the percentage of alpha Cronbach is 0.827 which means the highest value in credibility

Relative Importance Index:

The relative importance index is used to identify the ranking of the delaying factors and is considered as a statistical method used as analytical tools to get the ranking of delaying factors; the Relative importance index depends upon the Five Likert Scale. In the five Likert Scale: from one to five express with different cases (one: Strong Disagree, two: Disagree, three: Neutral, four: Agree, Five: Strongly Agree.

$$RII = \frac{\Sigma W}{A * N}$$

Where:

RII is Relative importance index.

W is Weighting is given to each other by the respondents (ranging from 1 to 5).

A is the highest weight (i.e. 5 in this case).

N is Total number of respondents.

The importance level of relative importance index can be categorized with different levels, Table (2) indicates the level of importance according to the value of the relative importance index.

IOP Conf. Series: Materials Science and Engineering **762** (2020) 012009 doi:10.1088/1757-899X/762/1/012009

RII values	Level of importance
$0.90 \leq \boldsymbol{RII} \leq \mathbf{1}$	Strongly important
$0.75 \leq \boldsymbol{RII} \leq \mathbf{0.89}$	Very important
$0.74 \leq \boldsymbol{RII} \leq \mathbf{0.60}$	important
$0.45 \leq \boldsymbol{RII} \leq \mathbf{0.59}$	Moderately important
$0.30 \leq \boldsymbol{RII} \leq \mathbf{0.44}$	Unimportant
$0.15 \leq \boldsymbol{RII} \leq \mathbf{0.29}$	Very unimportant
$0.0 \leq \boldsymbol{RII} \leq \mathbf{0.14}$	Strongly unimportant

From the current study the main delay factors effects on the extension of time of the construction project are the delaying of the supplies (Code of Factor NF5) with mean 4.2039 and RII = 0.840 (Level of importance = very important), A lot of change Orders (code of Factor NF 14) with mean 4.1845 RII = 0.8369 (Level of importance = very important), Delaying of issue the license and agreement to construct the project (code of Factor NF 1) with mean 4.1553 RII = 0.83106 (Level of importance = very important) and the poor management (code of Factor NF 17 with mean 4.106) RII = 0.82136 (Level of importance = very important), Table (2) and Fig (7) illustrates the relative importance index for the delaying factors in the construction projects in Egypt.

Table (2)

Ser	Code of Delay Factor	Description	Mean	Standard Deviation	RII	Importance Level
1	DF 1	Delaying of issue the license and agreement to construct the project	4.1553	1.02668	0.83106	Very important
2	DF 2	Lack of communication between different department inside the organization	3.9903	0.85744	0.79806	Very important
3	DF 3	Effects of absence of work authorization system	3.7864	0.9666	0.75728	Very important
4	DF 4	Absence of the communication requirement inside the organization	3.8835	0.94247	0.7767	Very important
5	DF 5	The delaying in the procurement process	4.1456	0.98427	0.82912	Very important
6	DF 6	The delaying of the supplies	4.2039	0.96365	0.84078	Very important
7	DF 7	the delaying of payment request for vendors and subcontractors	3.9903	1.00484	0.79806	Very important
8	DF 8	The delaying of the fee of the labor and incentives	3.6893	1.11169	0.73786	important
9	DF 9	Absence of justice in distribution the fees between the labors	3.3981	1.08772	0.67962	important
10	DF 10	Poor performance management of the subcontractors	3.7379	1.13726	0.74758	important
11	DF 11	Slow making decision	3.9223	0.98707	0.78446	Very important
12	DF 12	Hesitated and change in the decision making	3.8058	0.99072	0.76116	Very important
13	DF 13	careless the risk management factors in the projects	3.7087	1.07221	0.74174	important
14	DF 14	A lot of change Orders	4.1845	0.89385	0.8369	Very important

Ser	Code of Delay Factor	Description	Mean	Standard Deviation	RII	Importance Level
15	DF 15	Lack of Skilled workers	3.7476	0.96729	0.74952	important
16	DF 16	Absence of Inflation and the change in the prices of the materials, equipment and wages	3.7379	1.16283	0.74758	important
17	DF 17	the poor management	4.1068	0.87349	0.82136	Very important
18	DF 18	the weakness of the technical performance	3.8738	0.97699	0.77476	Very important
19	DF 19	the absence of communication outside the organization	3.835	0.85287	0.767	Very important
20	DF 20	Delaying of the deliverables at the working stages	3.9417	0.92699	0.78834	Very important
21	DF 21	the effects of type of leadership on the time and cost of the project	3.7184	0.98436	0.74368	important
22	DF 22	The effects of type of organization (centralized, decentralized)	3.5728	1.09904	0.71456	important

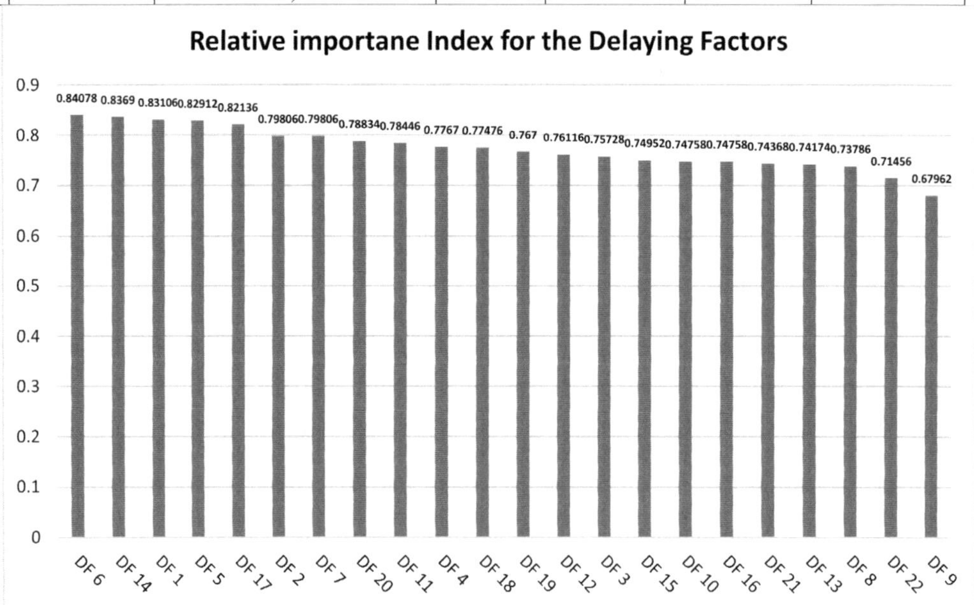

Fig 7

5.2 Factor Analysis

Factor analysis can be used to categorize the delay factor into groups, sorting the 22 delaying factors into seven groups and created the factors into modeling as a tool for management, the reliability of the modeling has been investigated by the communalities of each variable, Table (3) clarify the factor analysis of the delaying factors, it has been created by using SPSS software, the sample size has been investigated in our study is 103 samples,

Communalities indicate the common variance shared by factors with given variables. Higher communality indicated that a larger amount of the variance in the variable had been extracted by the factor solution. For better measurement of factor, analysis communalities should be 0.3 or greater. All the communalities are above 0.4, so that there are acceptance, In our study the delay factor analysis had a value bigger than 0.4 so that it can be reliable in the research to create the factor model, the parameters in factors analysis are Varimax Rotation and the value of KMO (Kaiser-Meyer-Olkin) = 0.664 which is greater than 0.5 means it is acceptance value, the results of factor analysis created 7-factor model, which is extracted depends on the eigen values more than 1, so that the factor models can name by the organization management (Factor model 1), The procurement process (Factor Model 2), the performance of worker and subcontractors and the wages incentive process (Factor Model 3), Make a decision (factor model 4), Communication process (Factor Model 5), The inflation and delaying in payment request, Fairless in the fee (Factor Model 6), and License and the work authorization system (Factor model 7) respectively for factor model 1 & 2 &3 &4 &5 &6 & 7, it can be shown in Fig (8), and the factor analysis created seven core models to represent 22 delay factors

Table (3)

Code of Delay Factor	Factor 1	Factor 2	Factor 3	Factor 4	Factor 5	Factor 6	Factor 7
DF21	0.732						
DF18	0.700						
DF17	0.642						
DF22	0.617						
DF14	0.596						
DF13	0.519						
DF6		0.844					
DF5		0.815					
DF15			0.809				
DF10			0.706				
DF8			0.459				
DF20			0.436				
DF12				0.831			
DF11				0.772			
DF2					0.819		
DF4					0.653		
DF19					0.563		
DF9						0.608	
DF7						0.569	
DF16						0.493	
DF1							0.832
DF3							0.790

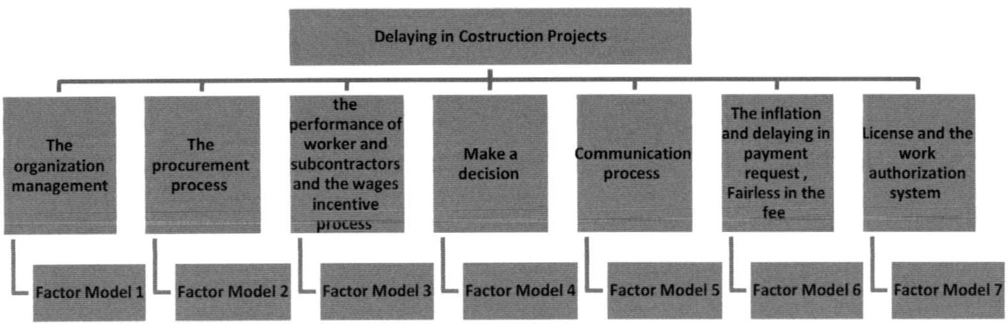

Fig (8)

5.3 Fault Tree analysis:

The fault tree of the delaying factors constructed as shown Fig (9), it composed the top undesired event (delaying of the construction projects) and the intermediate events composed into seven Events which link to the basic events of the delaying factors, the intermediate events of the delaying factors submitted from factor analysis modeling, they compose to the organization management, the procurement process, the performance of worker and subcontractors and the wages incentive process, make a decision, Communication process, The inflation and delaying in payment request, Fairless in the fee, and License and the work authorization system.

Each intermediate event connect to the basic events by logical gates, where the organization management connects to the basic events the effects of type of leadership on the time and cost of the project, the weakness of the technical performance, the poor management. The effects of type of organization (centralized, decentralized). A lot of change Orders, and Carless the risk management factors in the projects.

Intermediate event (the procurement process) connects to the basic events (the delaying of supplies and the delaying in the procurement process)

Intermediate event (the performance of worker and subcontractors and the wages incentive process) connects to the basic events (Lack of Skilled workers, Poor performance management of the subcontractors, the delaying of the fee of the labor and incentives, and Delaying of the deliverables at the working stages)

The intermediate event (Make a decision) connects to the basic events (Hesitated and change in the decision making, slow making decision)

The intermediate event (Communication process) connects to the basic events (Lack of communication between different departments inside the organization, absence of the communication requirement inside the organization and the absence of communication outside the organization)

The intermediate event (The inflation and delaying in payment request, Fairless in the fee) connects to the basic events (Absence of justice in distribution the fees between the labors, the delaying of payment request for vendors and subcontractors, and absence of Inflation and the change in the prices of the materials, Equipment, and wages)

The intermediate event (License and the work authorization system) connects to the basic events (Delaying of the issue the license and agreement to construct the project)

Fig (9) illustrate the fault tree analysis with all components from the top event to the basic events and the connections between all events with the logical gates

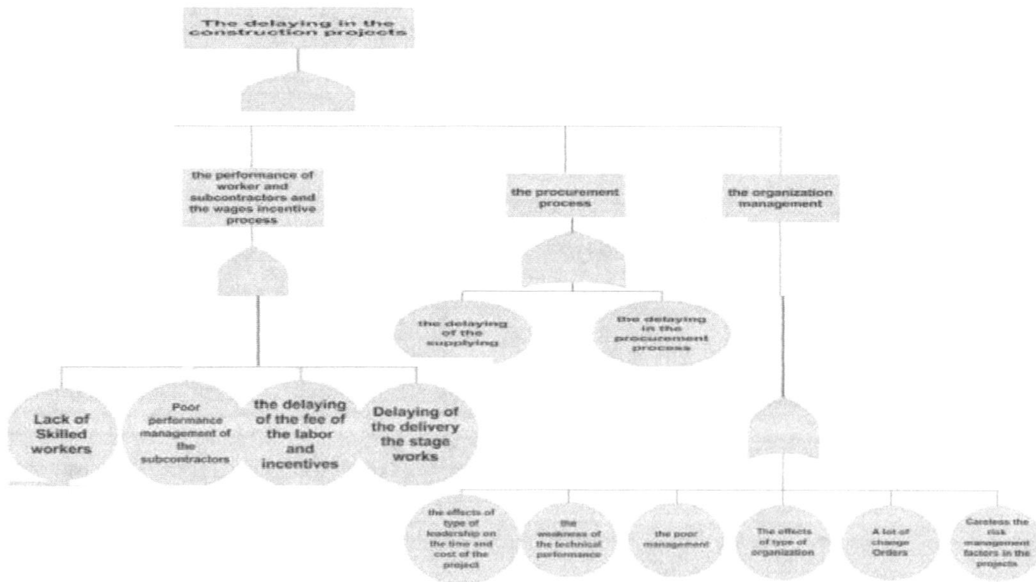

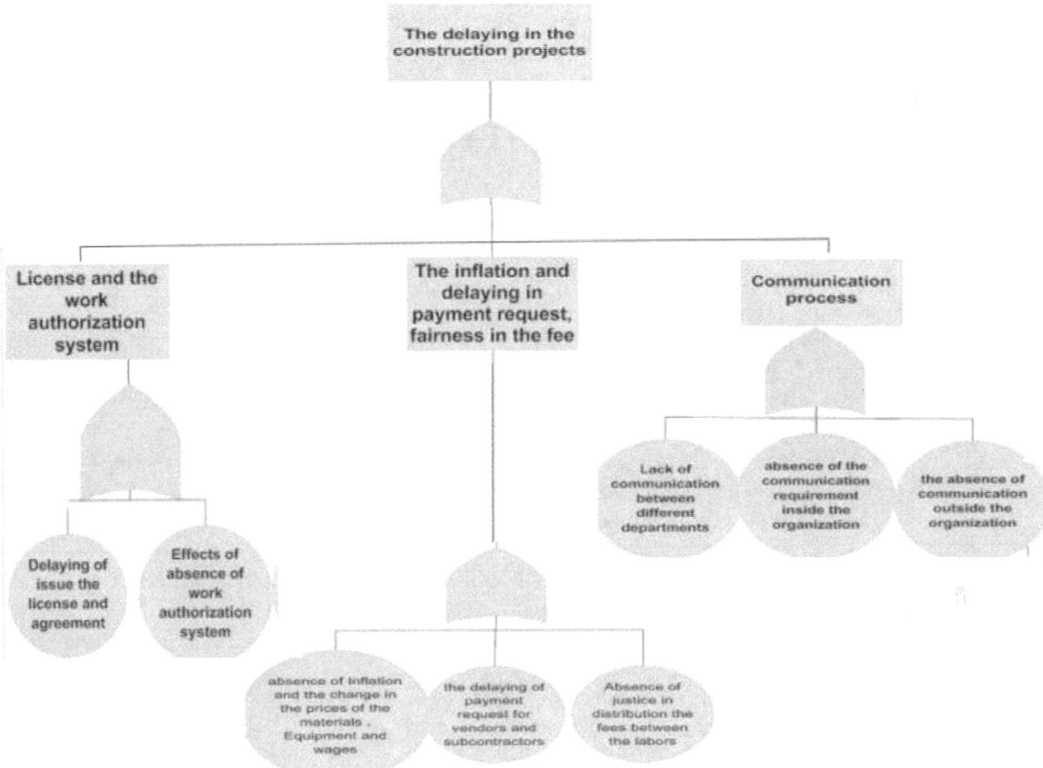

Fig (9)

6. Conclusion:

Using fault tree analysis to determine the main delaying causes for the construction projects, it can identify the primary causes of delaying (the organization management, the procurement process, the performance of worker and subcontractors and the wages incentive process, make a decision, Communication process, The inflation and delaying in payment request, Fairless in the fee, and License and the work authorization system).

Due to the statistical analysis: the relative importance index for the delaying of the supplying has the biggest value for RII (RII = 0.84078) , and then A lot of change Orders with RII = 0.8369 and then Delaying of the issue the license and agreement to construct the project with RII = 0.83106,

Using the factor analysis in the creation of the modeling of factor analysis, FAM composes seven-factor models with the main causes of delaying

In the final study: the effects of delaying of supplies is the most significant problems and to solve these problems or avoiding it can be achieved by following the process of management of procurements, Measure and evaluate the vendors performance are important for solving the problem of extension time in the project so that it should create the performance measurement for suppliers which contains the communication system to easy the contact with the vendors to solve the shortcoming early, recommendation for the organizations is to start to measure key suppliers delivery performance and support the vendors to improve the delivery performance

Many change orders are the other big problems lead to extension in time in the construction project, the change orders have been caused by the errors or omission in the project scope or missing the details, misinterpretation in the drawings or other reasons and to prevent the common problems, it can achieve by review the contract documents and specification before starting the construction projects and confirm that the owner should provide an accurate information during the project life cycle

References

[1] Aibinu, A.A. and Jagboro, G.O. (2002). *The effects of construction delays on project delivery in Nigerian*. International Journal of Project Management, **20** 593–599.
[2] Assaf, S.A. and Al-Hejji, S. (2006). *Causes of delay in large construction projects*. International Journal of Project Management, **24.4** 349–357
[3] Faridi, A.S. and El-Sayegh, S.M. (2006). *Significant factors causing delay in the UAE construction industry*. Construction Management and Economics
[4] Fallahnejad, M.H. (2013). *Delay causes in Iran gas pipeline projects*. International Journal of Project Management **31** (2013) 136–146.
[5] O'Brien JJ. *Construction delay: responsibilities, risk and litigation*. Boston, US: Cahners Books International, Inc.;1976
[6] Doloi H, Sawhney A, Iyer KC, Rentala S. *Int J Project Manager* 2012 Analyzing factors affecting delays in Indian construction projects **30.4** 479–89.
[7] Sweis G, Sweis R, Abu Hammad A, Shboul *2008 Int J Project Manager* A. Delays in construction projects the case of Jordan **26.6** 665–74.
[8] Hamzah N, Khoiry MA, Arshad I, Tawil NM, Ani Che AI. 2011 *Procedia Eng* Cause of construction delay – theoretical framework **20** 490–5
[9] Pickavance K. 2000 *Delay and disruption in construction contracts*. 2nd ed. London, UK: Sweet & Maxwell.
[10] Bordoli DW, Baldwin AN. 1998 *J Constr Manage Eng* A methodology for assessing construction project delays **16.3** 327–37
[11] Giridhar P, Ramesh K. 1998 *Effective management of Turnkey projects*. Aace Transactions, PM7-PM11.
[12] Mostafa Abd Elrazek, Yasser Abdel Shafy Gamal. 2019 *sciencepub.net/new york* Study Of High Impact Factors Effecting on Delaying Construction Projects in Egypt. N Y Sci J 2019;**12(2):**4-15]. ISSN 1554-0200 (print); ISSN 2375-723X (online). http://www.sciencepub.net/newyork.2.doi:10.7537/marsnys120219.02
[13] Hair, J.F., Black, W.C., Babin, B.J., and Anderson, R.E. (2010). *Multivariate data analysis*, 7th Ed., Pearson Education, Upper Saddle River, NJ.
[14] Battikha M G. 2008 *Journal of Construction Engineering and Management* Reasoning mechanism for construction non-conformance root-cause analysis. **134.4** 280-288.
[15] Andreas Lindhe, L.R., T. Norberg and O. Bergstedt, 2009 *Water research* Fault tree analysis for integrated and probabilistic risk analysis of drinking water systems. **43** 1641-1653.

[16] P.V. Suresh, A.K.B.a.V.V. 1996 Uncertainty in fault tree analysis: A fuzzy approach. Fuzzy sets and Systems **83** 135-141

AUTHOR INDEX